Construction Documents using SketchUp Pro & LayOut

by Paul Lee

Updated 2013 Edition

Complete Course
Create 2D construction information from your Model.

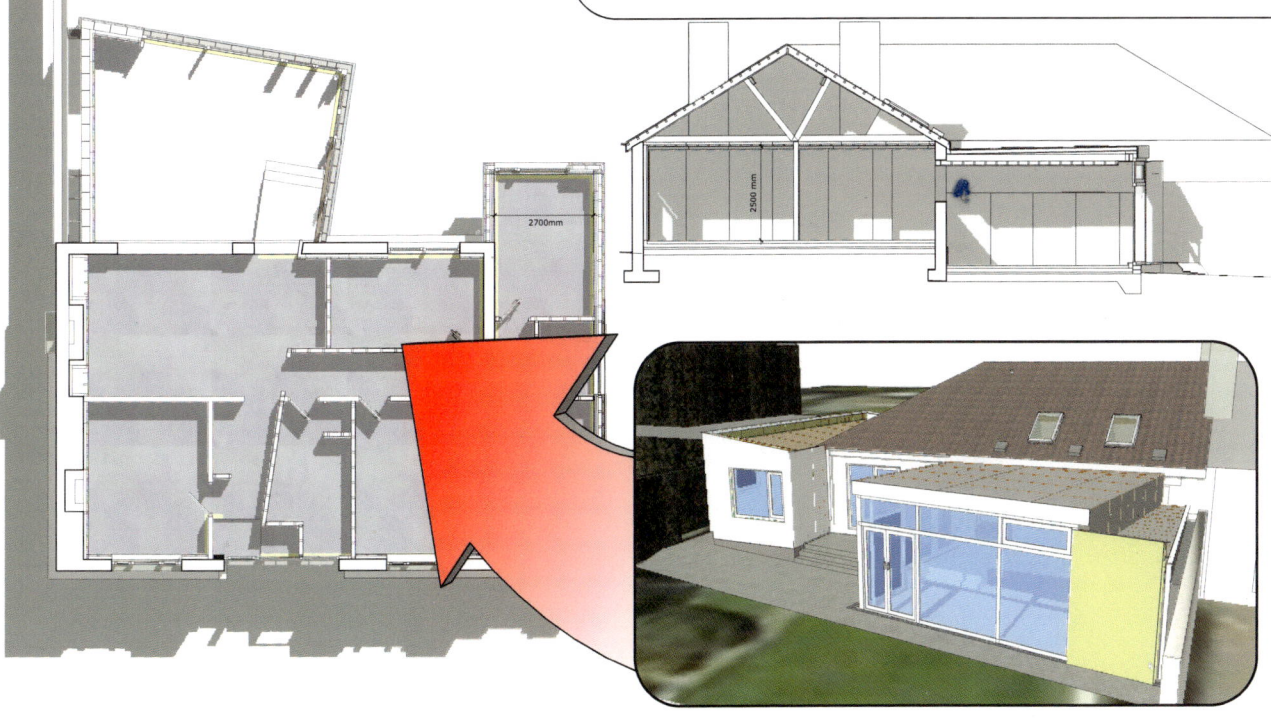

SketchUp2BIM™ Series

Construction Documents in Trimble SketchUp™ Pro by Paul Lee- *A methodology for creating technical 2D drawings* -©Paul Lee, 2013- www.viewsion.ie

Creating Construction Documents in Trimble SketchUp Pro

Introduction

For the past quarter century the construction industry has transitioned from the drawing board to the computer as the machine of choice for the creation of construction drawings.

This was an advancement in certain respects but perhaps not as great a leap forward as we might like to think. For all it's clumsiness, the drawing board provided the space for a quick overview of all the drawings together in context. The computer screen did away with this ability to assess drawings globally, instead providing merely an isolated view of a single drawing at a time.

Some would argue that this has created a decline of standards of design and construction. However, the new 3D paradigm is where a model contains all the information in one entity. This allows the practitioner to view all the pertinent project information in a single window. No longer do drawings need to be manually coordinated, nor do elevations ever need to be drawn: The model takes care of all these things.

SketchUp2BIM™ Series — Construction Documents in Trimble SketchUp™ Pro by Paul Lee- *A methodology for creating technical 2D drawings*- ©Paul Lee, 2013- www.viewsion.ie

Creating Construction Documents in Trimble SketchUp Pro

Conclusion

For the first time the computer can really deliver on the promise of an efficient drawing tool by distilling 2D construction information from a fully coordinated accurate 3D model.

However, the means of achieving 2D construction data from a SketchUp model has not been made clear- until now.

This manual is the first of four which delivers the know-how to produce full construction drawings in SketchUp Pro. This course outlines an expert workflow that combines:

- + Section Cuts
- + Grouping
- + Scenes
- + Layers
- + Styles

to produce a professional document containing "traditional" plans, sections elevations and details.

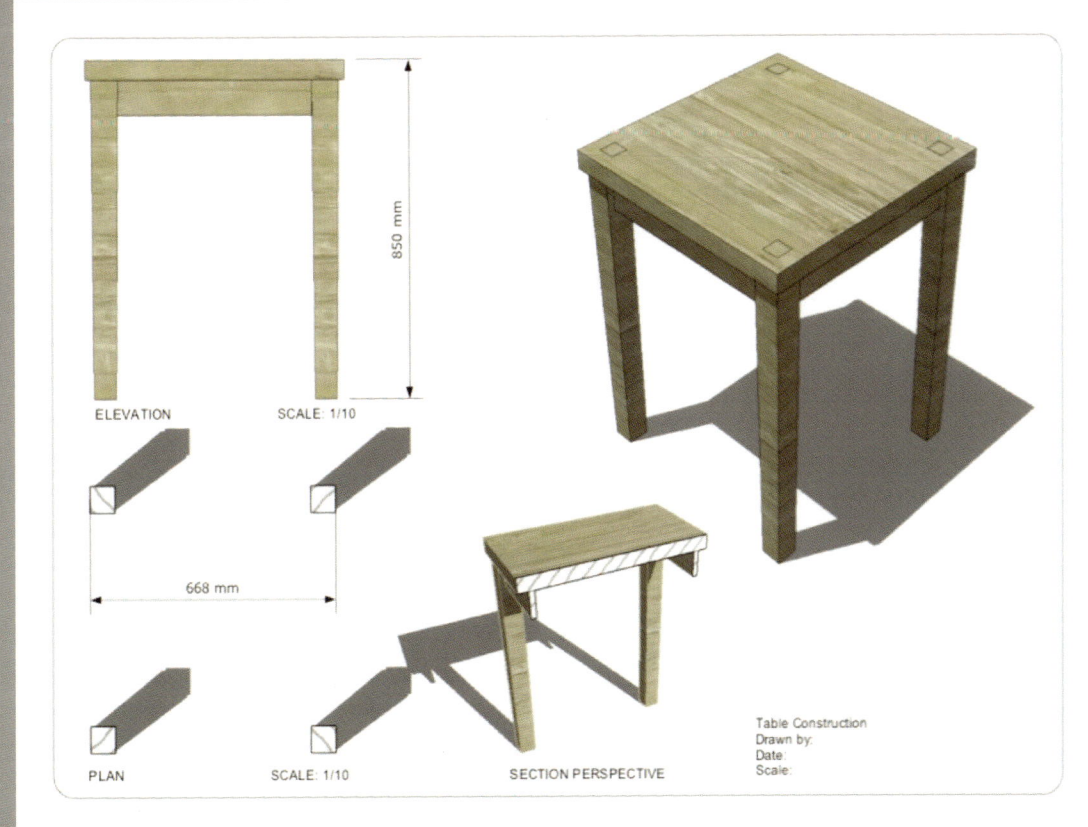

Target Drawing for this Manual

Download and open the Course Models:

http://www.viewsion.ie/#!publications

Contents:

To be used with Parts 1 - 3

1. SketchUp for Construction Documentation- Table
2. SketchUp for Construction Documentation- Table_Complete Model
3. Construction Documents in SU Pro_Table_Sample File

To be used with Part 4

4. Construction Documents in SketchUp Pro by Paul Lee 2012
5. Construction Documents in SketchUp Pro by Paul Lee 2012_Blank Model
6. Construction Drawings in SketchUp Pro by Paul Lee_ Part 4_Target Drawing

SketchUp2BIM™ Series — Construction Documents in Trimble SketchUp™ Pro by Paul Lee- *A methodology for creating technical 2D drawings*- ©Paul Lee, 2013- www.viewsion.ie

Creating Construction Documents in Trimble SketchUp Pro

Course Outline

The course is divided into four parts:

Part 1   Creating the 2D construction information.
Part 2   Setting up the Scenes in preparation for importing into LayOut.
Part 3   LayOut: Creating the construction information and Exporting to PDF.
Part 4   Structuring your Model and Recap Exercise.

Part 1

SketchUp allows you to create an instant section "drawing" of your model, called a "wireframe" Part 1 of this course shows how a wireframe can be displayed in front of the cut model (see below).

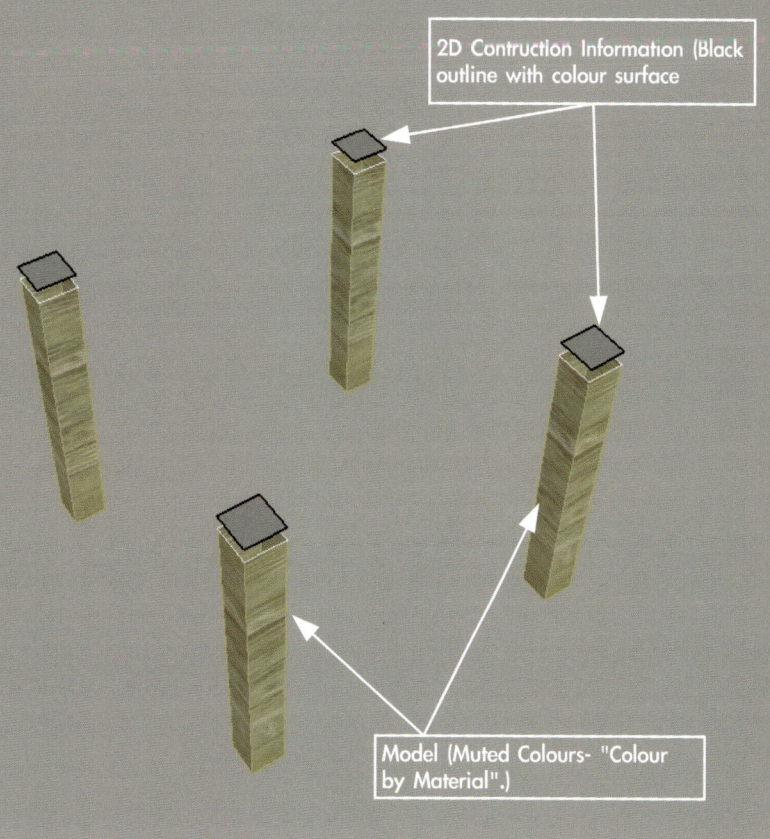

2D Contruction Information (Black outline with colour surface

Model (Muted Colours- "Colour by Material".)

Start

✓ Open SketchUp Pro.
✓ Make sure the Large Tool Set is visible (see below).

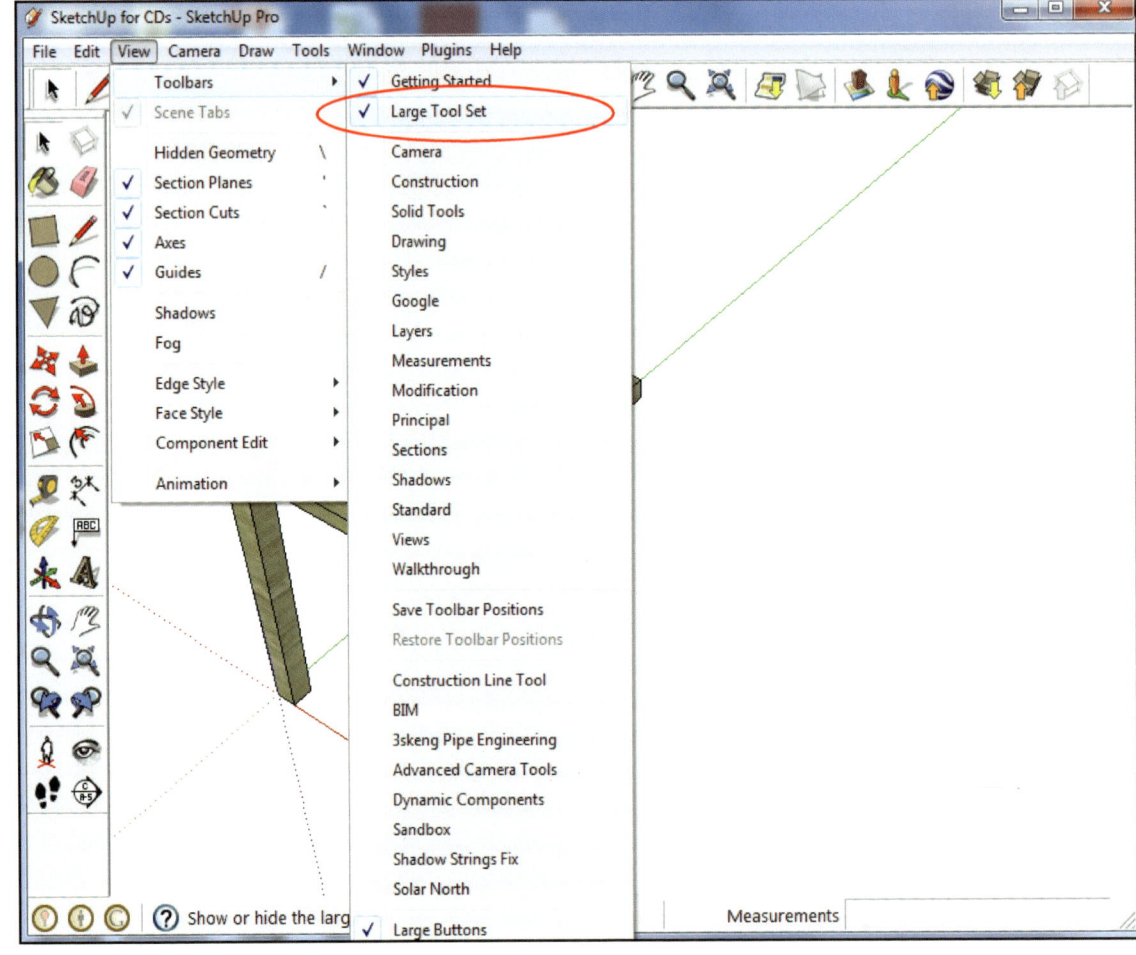

The Section Plane

✓ Click on the Section Plane button

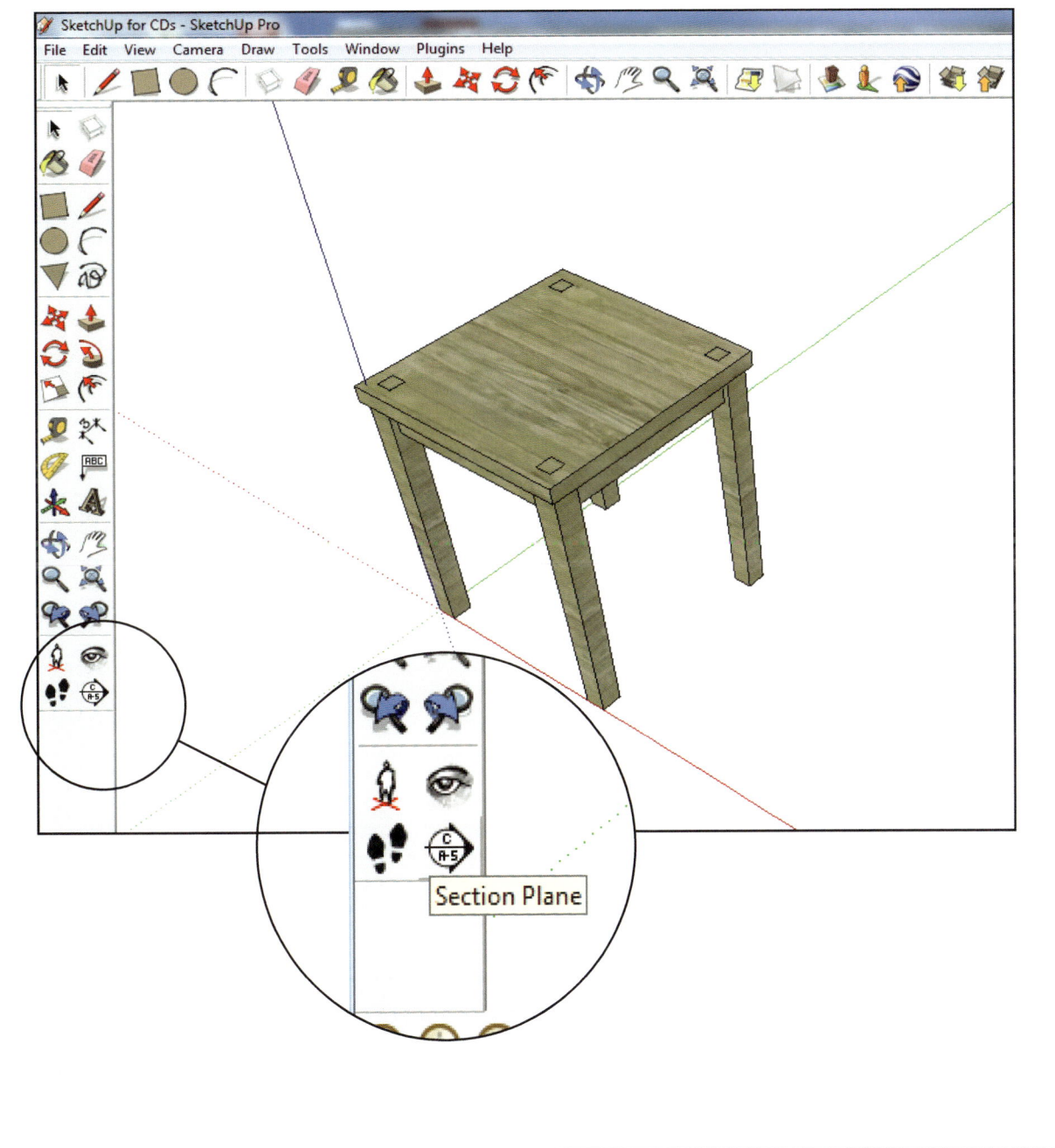

SketchUp2BIM™ Series — Construction Documents in Trimble SketchUp™ Pro by Paul Lee - *A methodology for creating technical 2D drawings* - ©Paul Lee, 2013 - www.viewsion.ie

Create the Section (1)

✓ When applying the section plane, hold down the SHIFT key to lock it into a plane

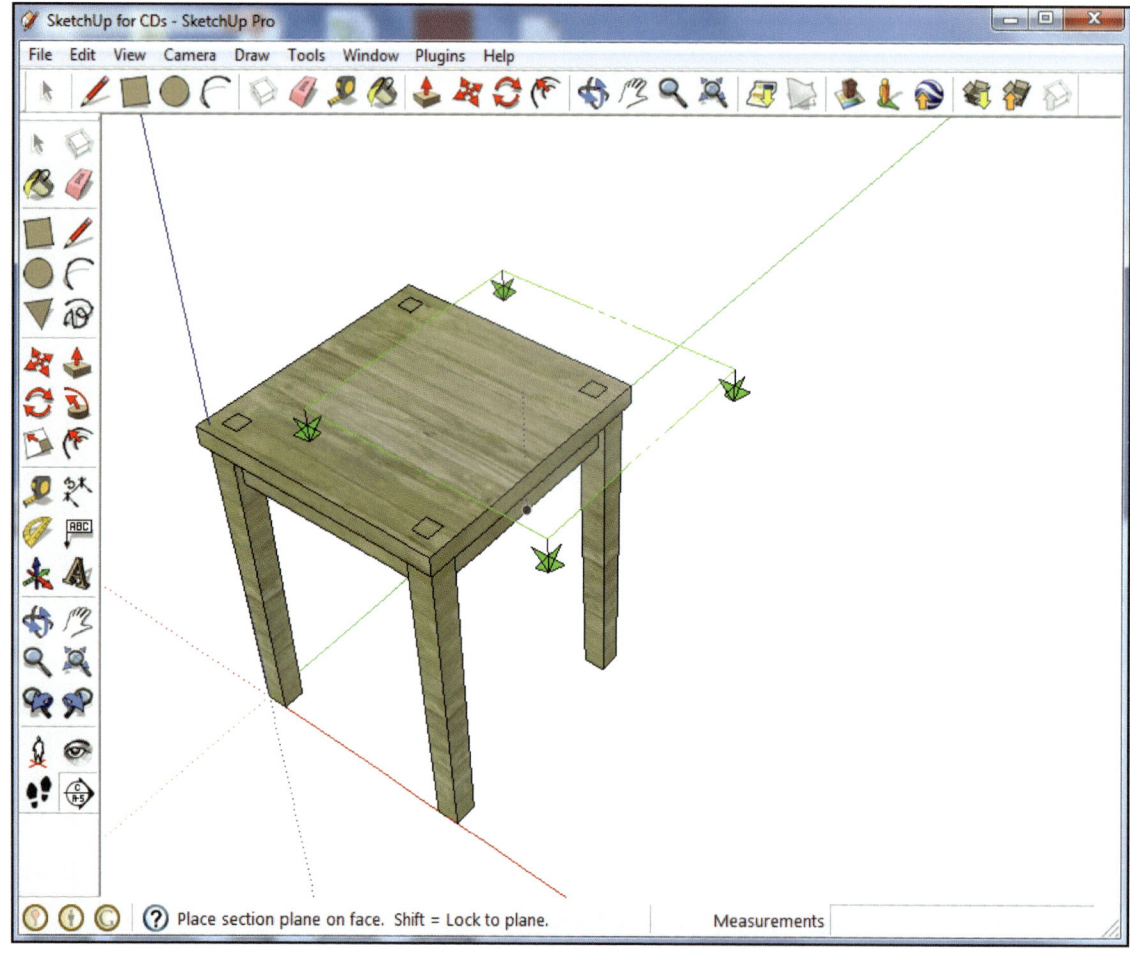

Create the Section (2)

✓ When the section plane is in place, left-click the mouse button to create the cut

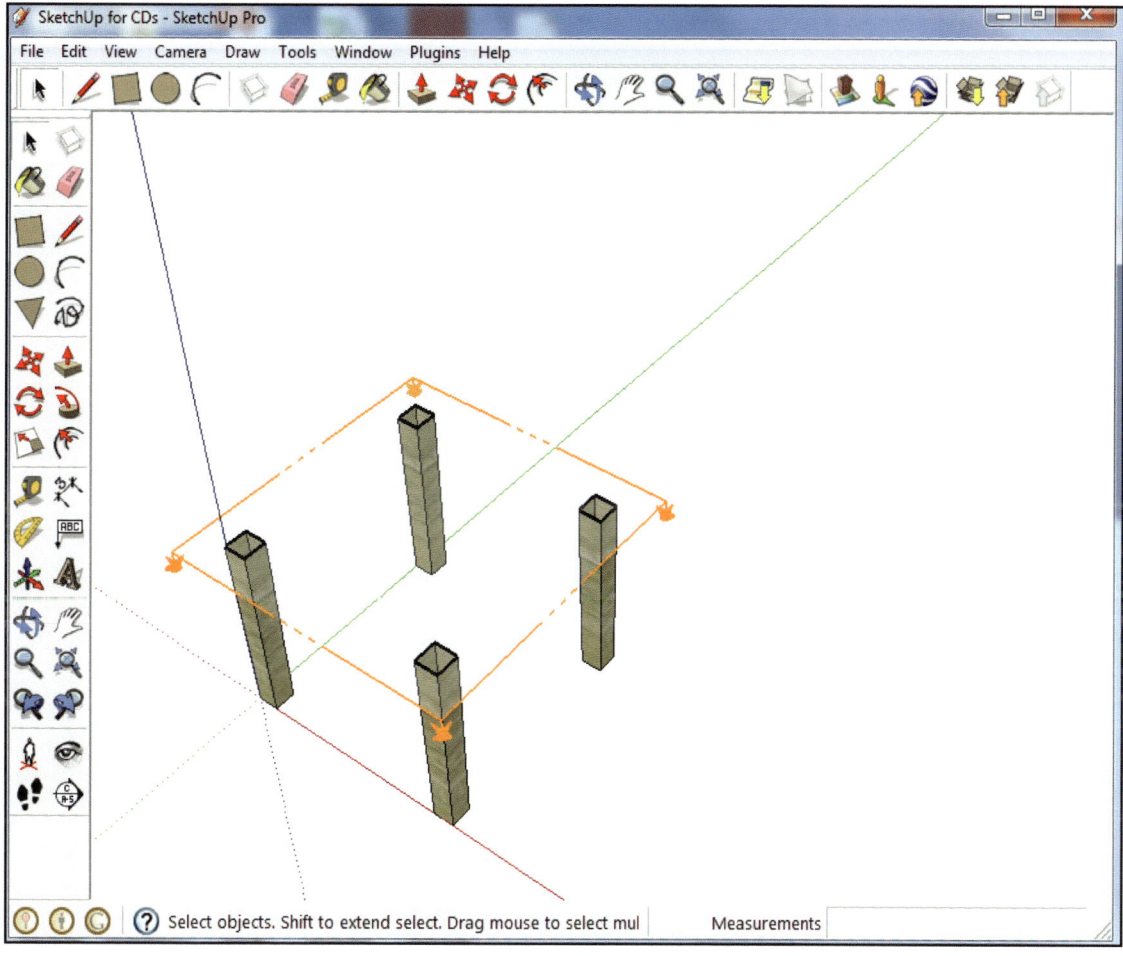

Open the Outliner Window

✓ Open Outliner as indicated below

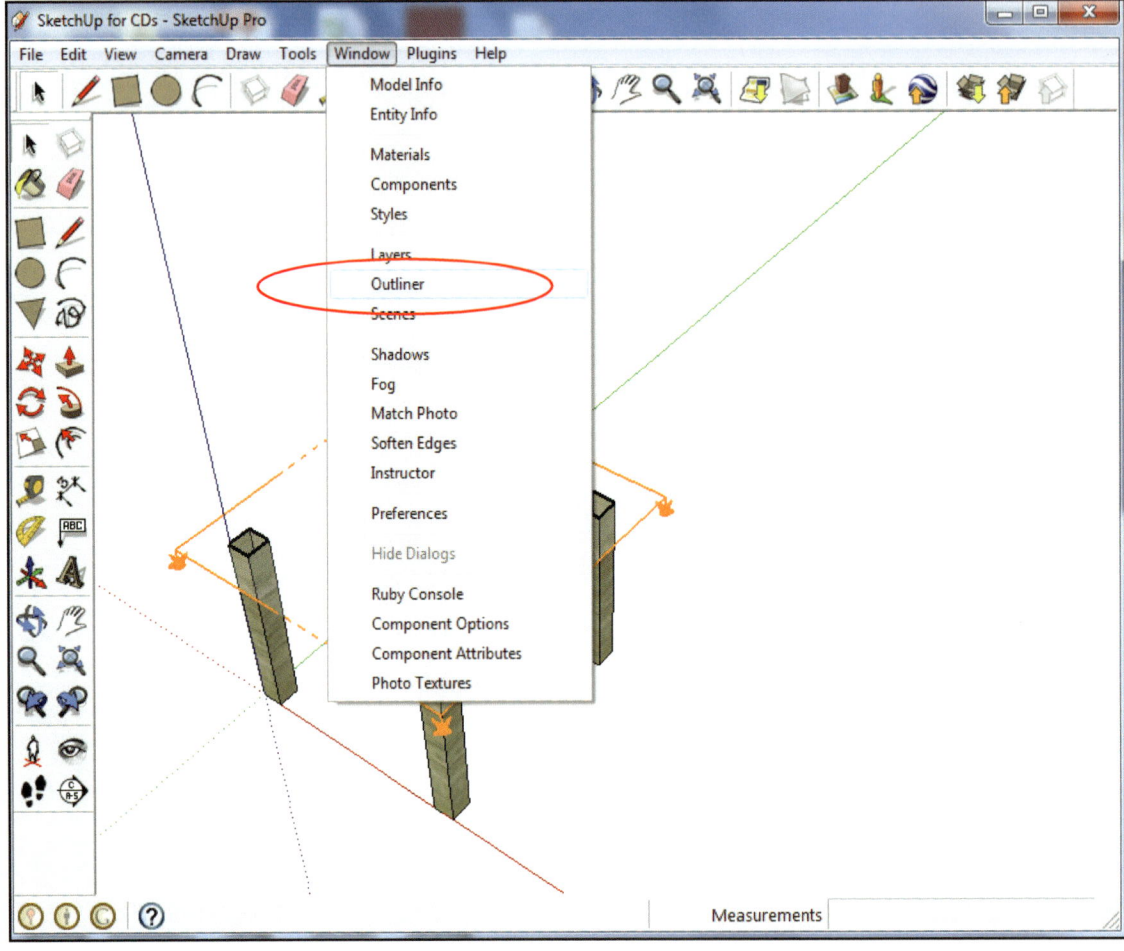

Create the Wire Frame

✓ Right-click on the section cut and select "Create Group from Slice"

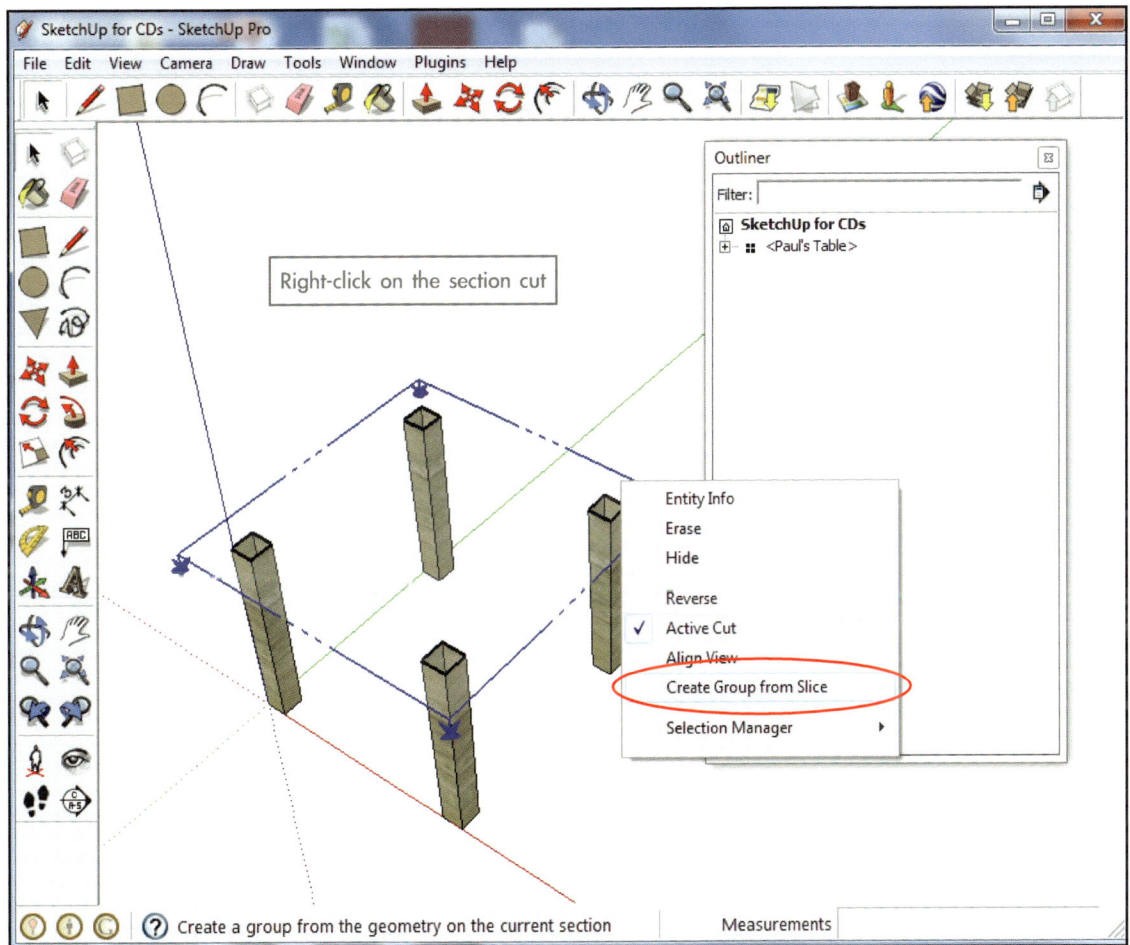

This action creates a 2D wireframe entity that will become the basis for the construction detail.

Outliner

✓ Notice that a new Group has appeared in Outliner. This is the 2D Wireframe.

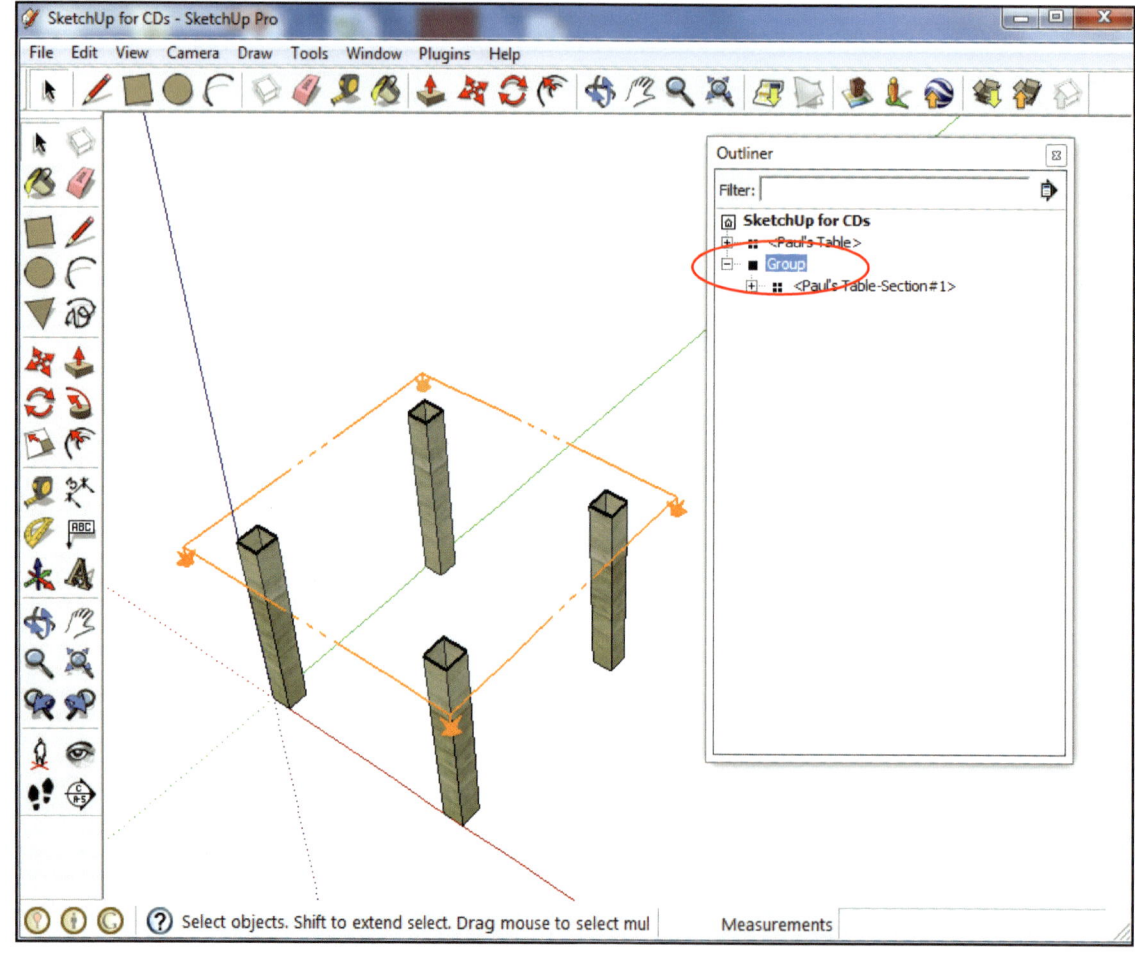

Group the Model & Section Plane together

- ✓ Notice that the Section Plane does NOT appear in Outliner.
- ✓ Select the Model and the Section Plane together.
- ✓ **Make sure NOT to select the 2D Wireframe!**

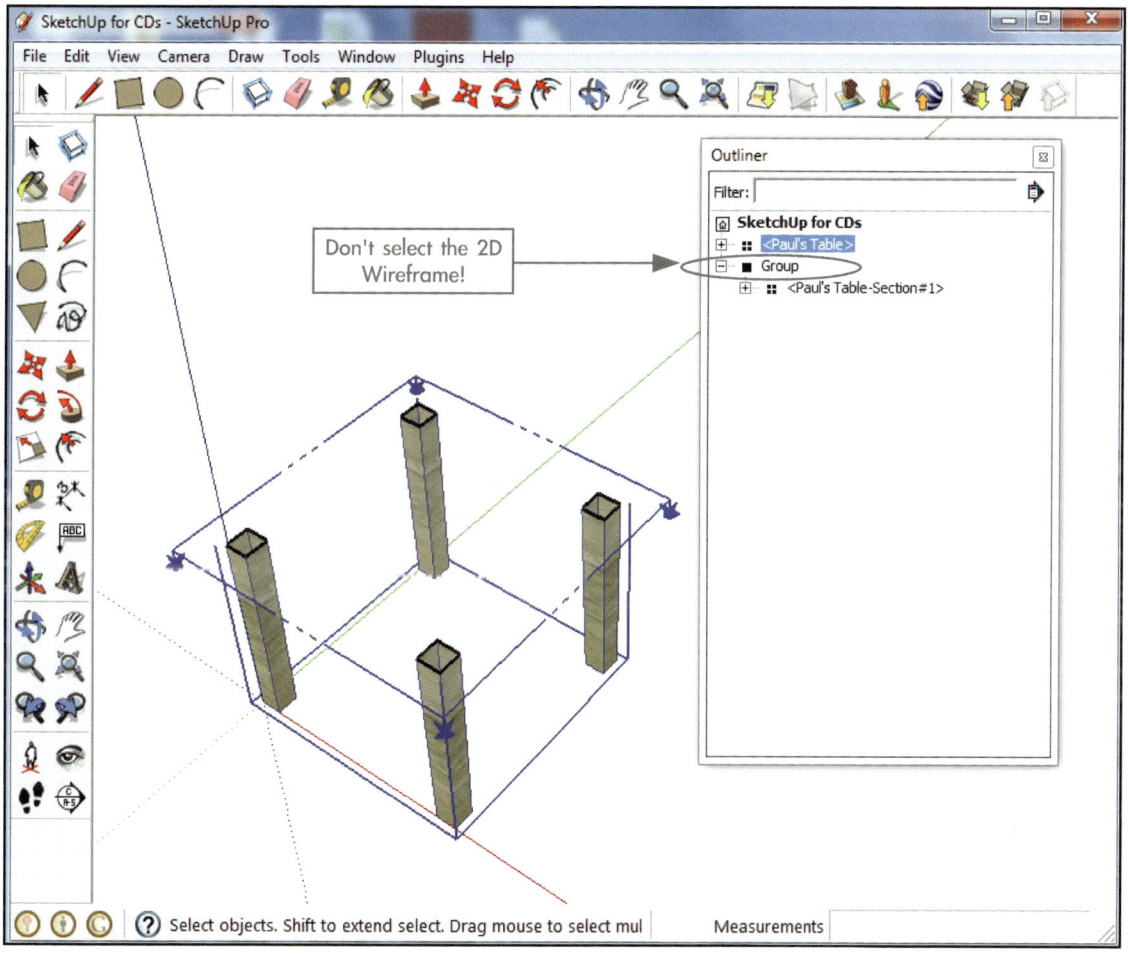

Make a Group of the Model & Section Plane

✓ Right-click on the selection and pick "Make Group"

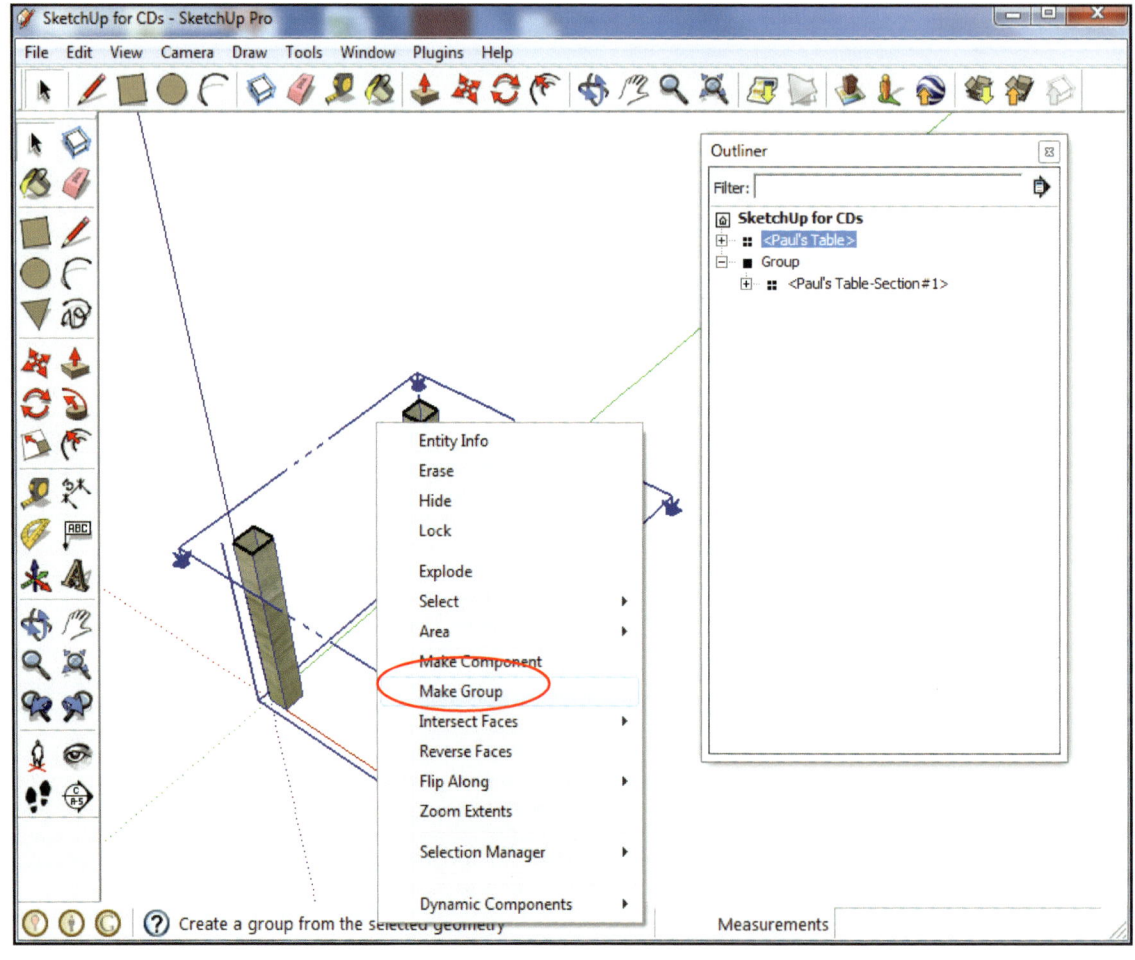

Group Formed

- ✓ The new Group has now been formed
- ✓ Notice that the 2D Section Plane now lies OUTSIDE the Group

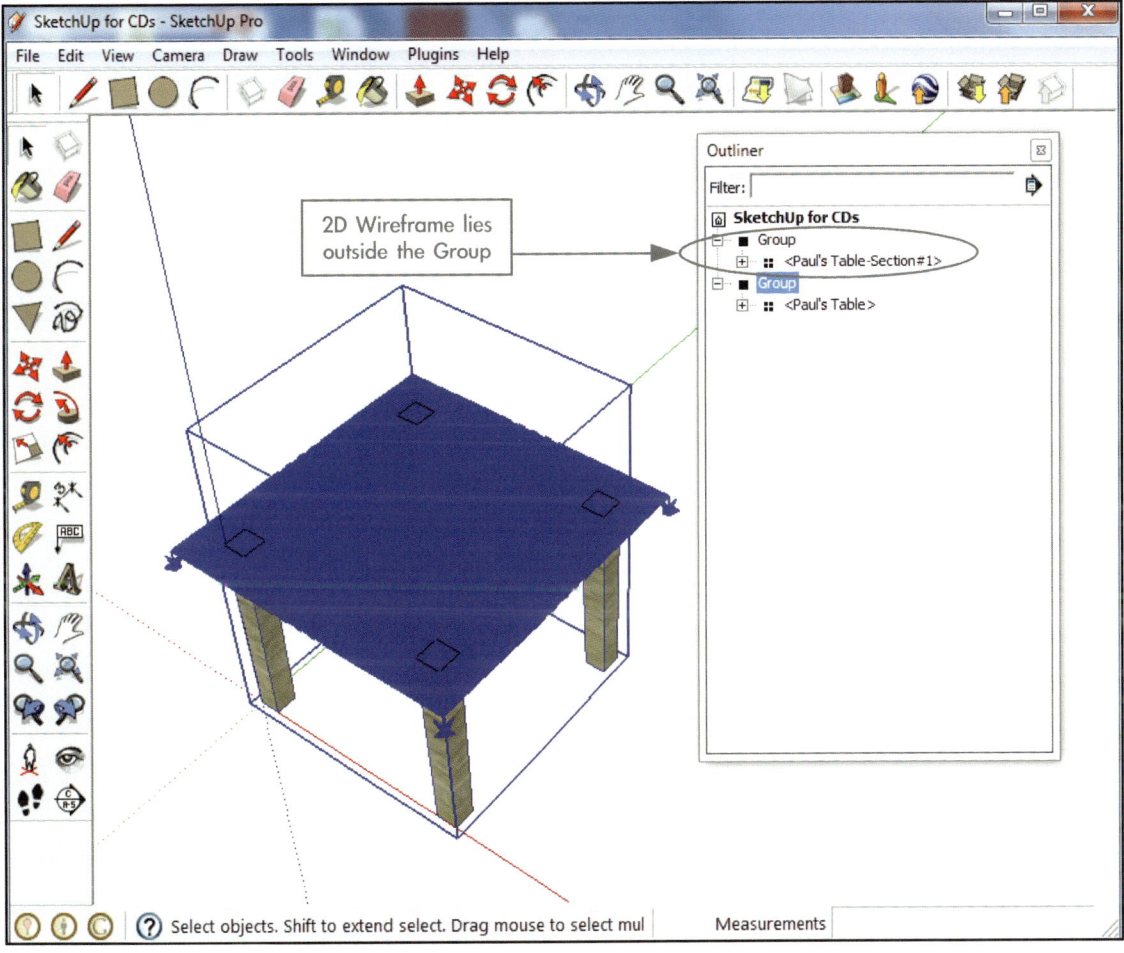

Rename the Group (1)

✓ Right-click on the Group and select "rename"

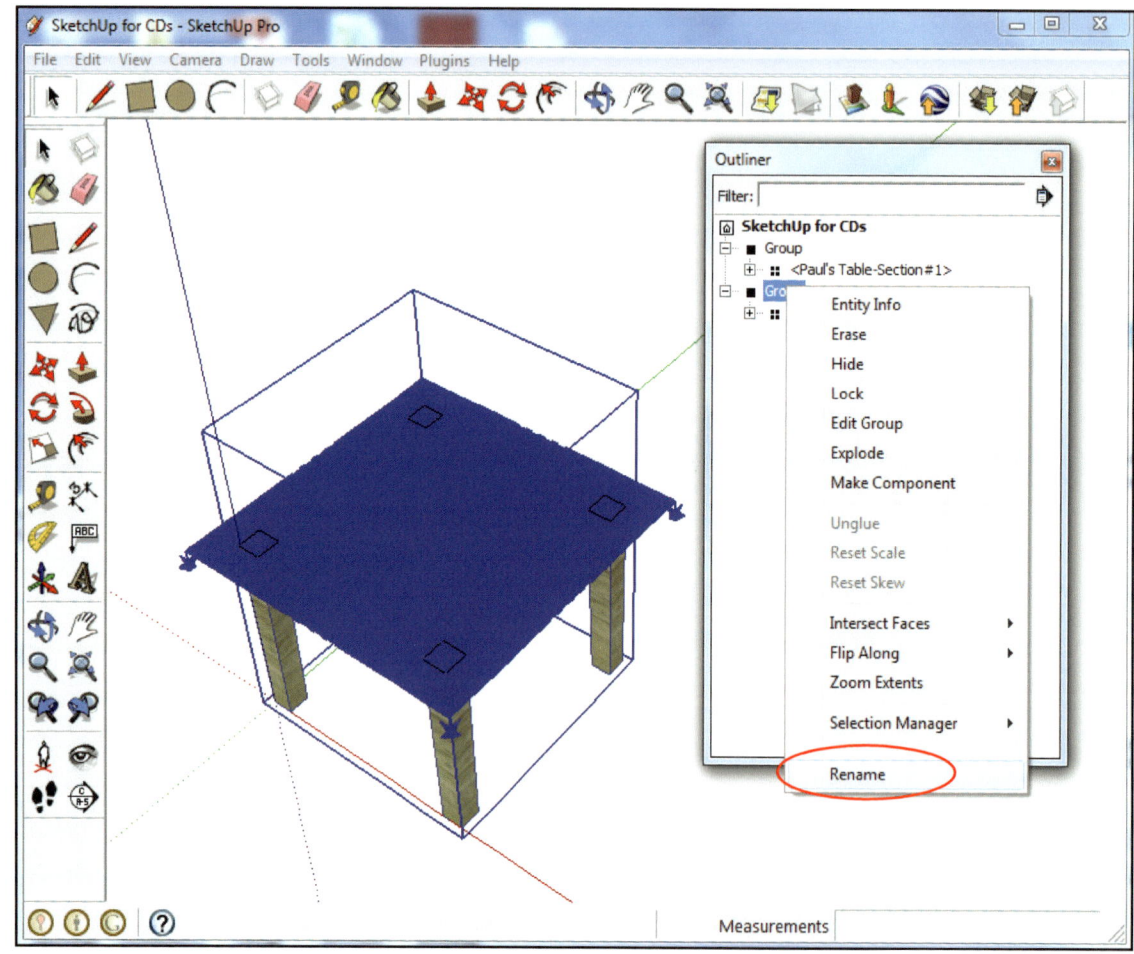

Rename the Group (2)

✓ Rename the Group as "Main Model"

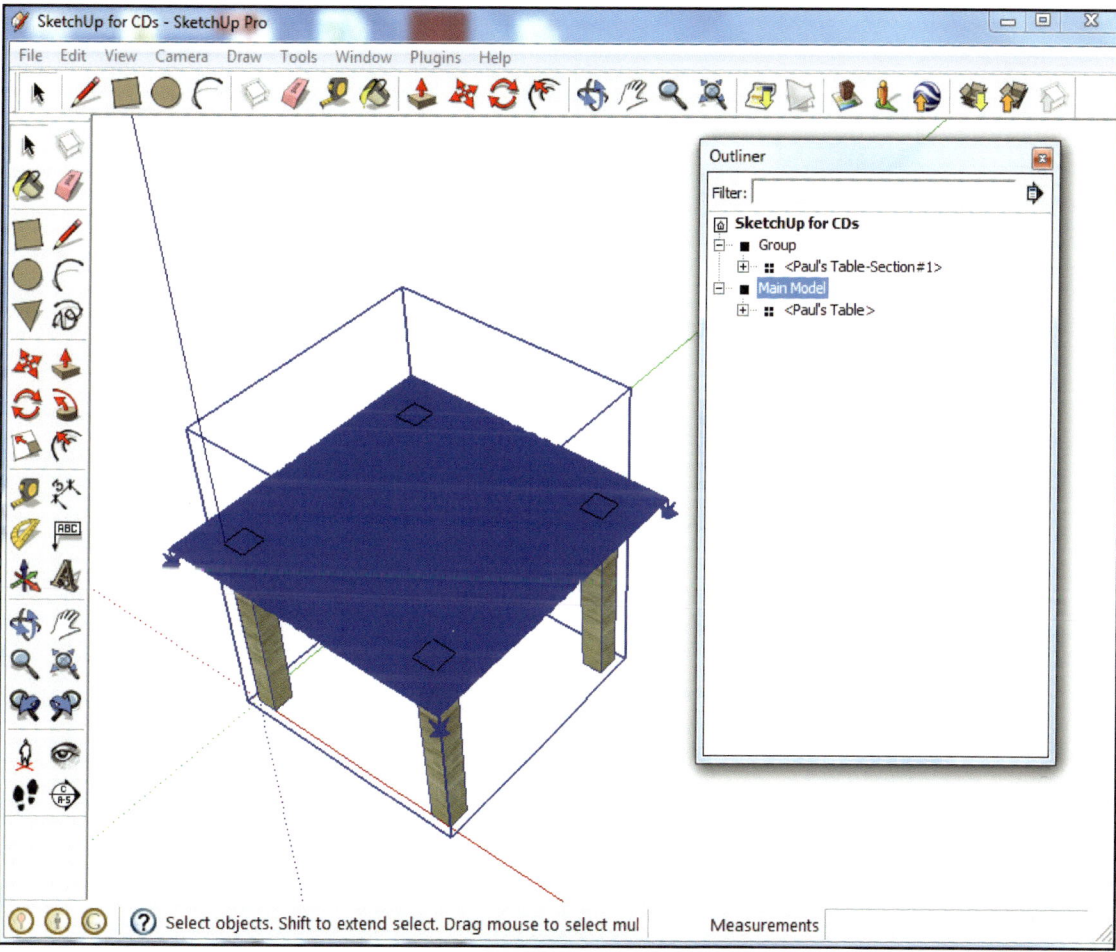

Open the Sections Toolbar

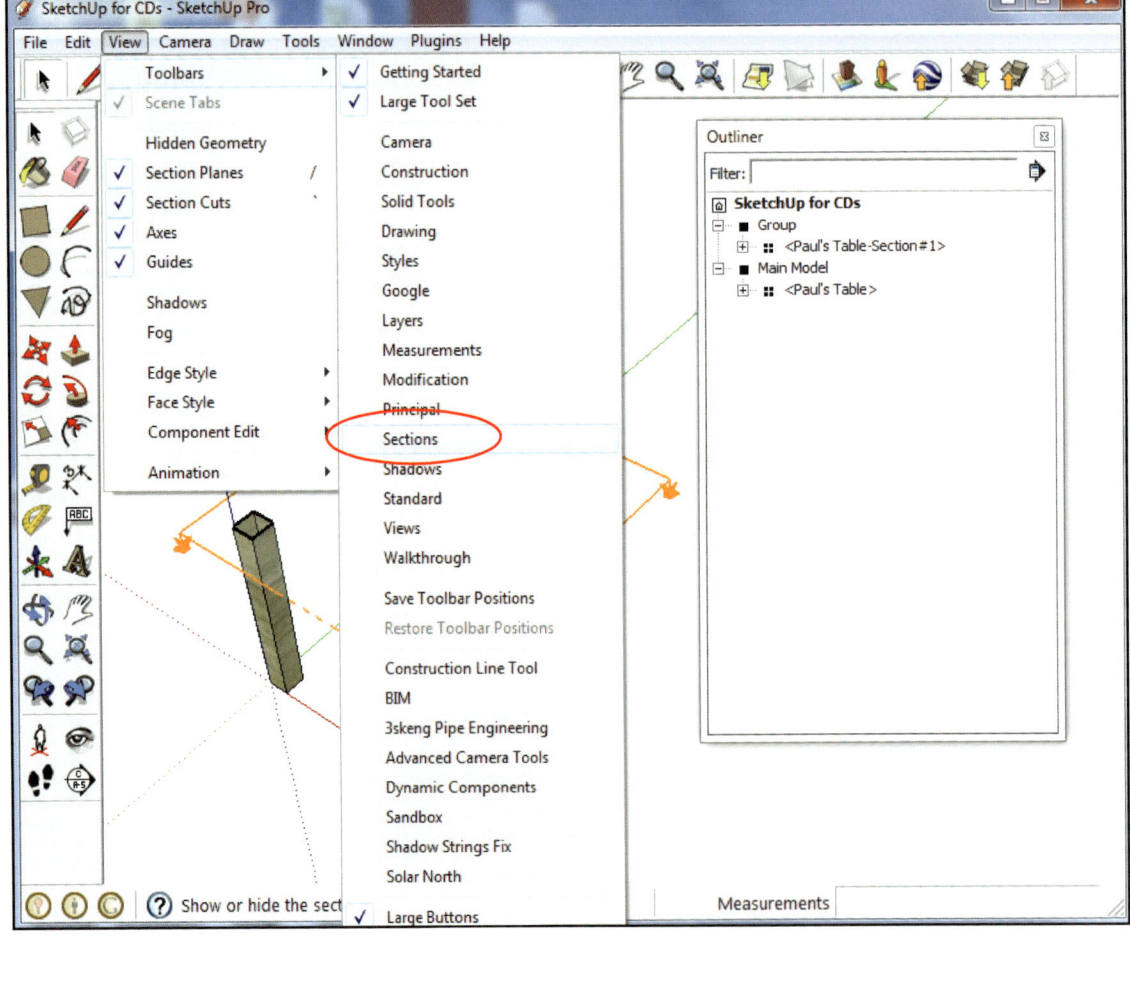

The Sections toolbar pops up

Activate/ Deactivate Section Cut

✓ Notice how the Section Activate/ Deactivate Button Works by clicking on it a few times (see below).

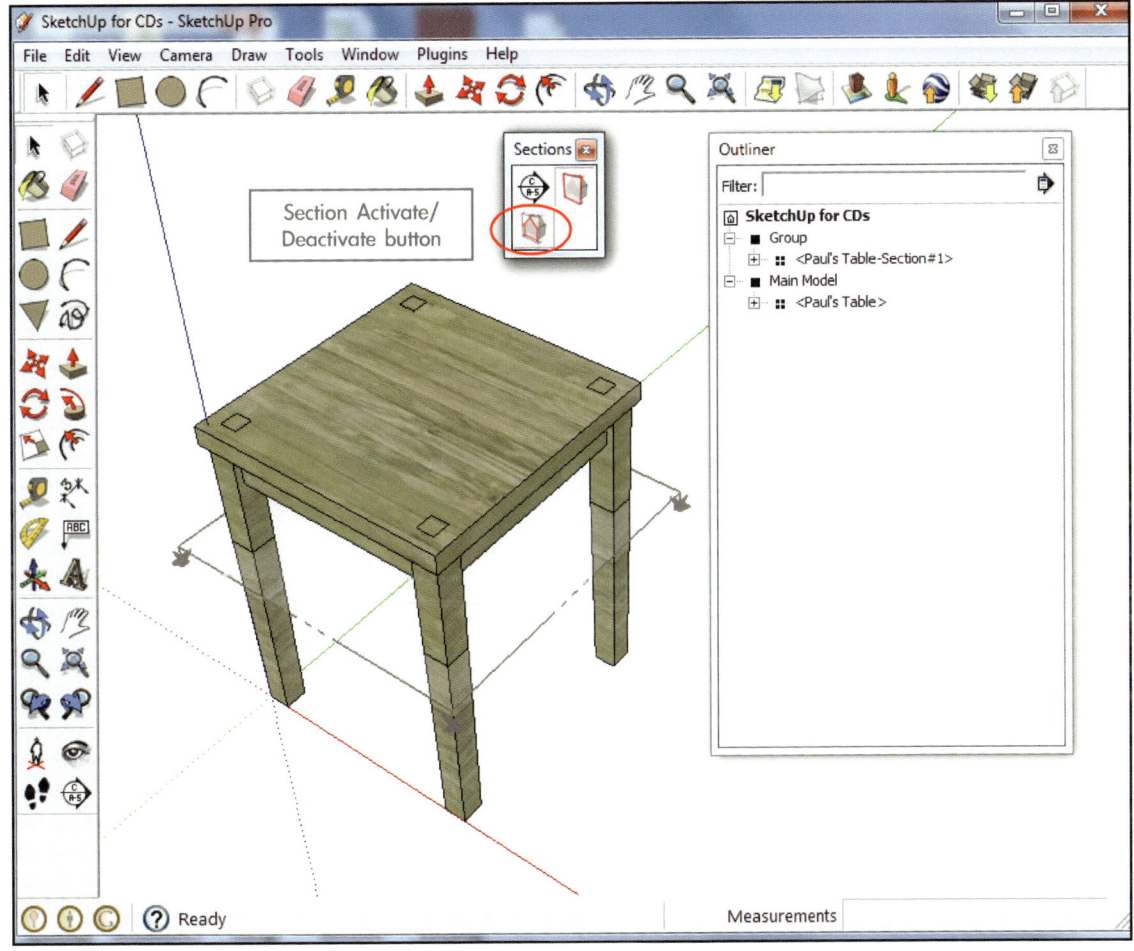

Activate/ Deactivate visibility of Section Plane

✓ Notice how the Section Plane appears and disappears by clicking on the Section Plane visibility button.

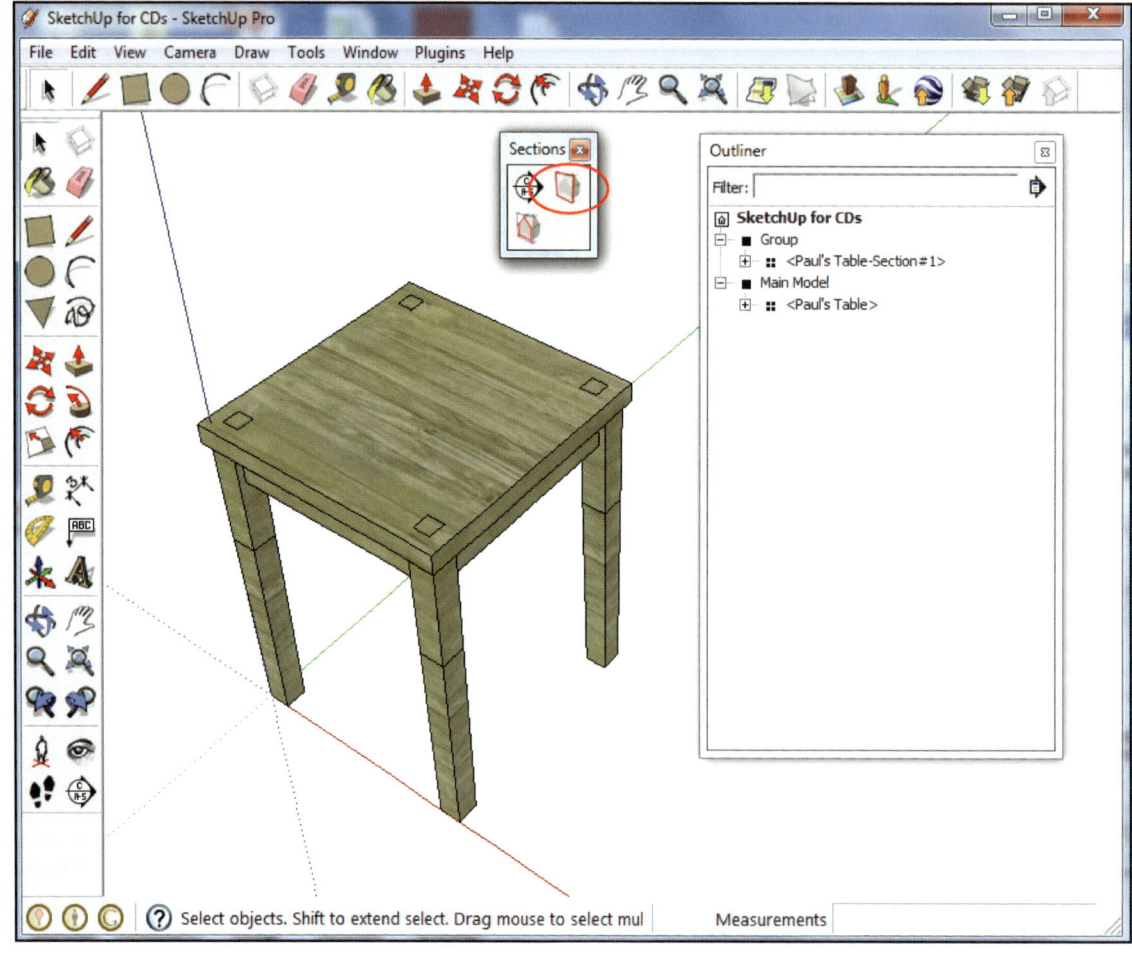

Deactivate the Section Plane

✓ Use the command button in the Sections Dialogue as highlighted below.

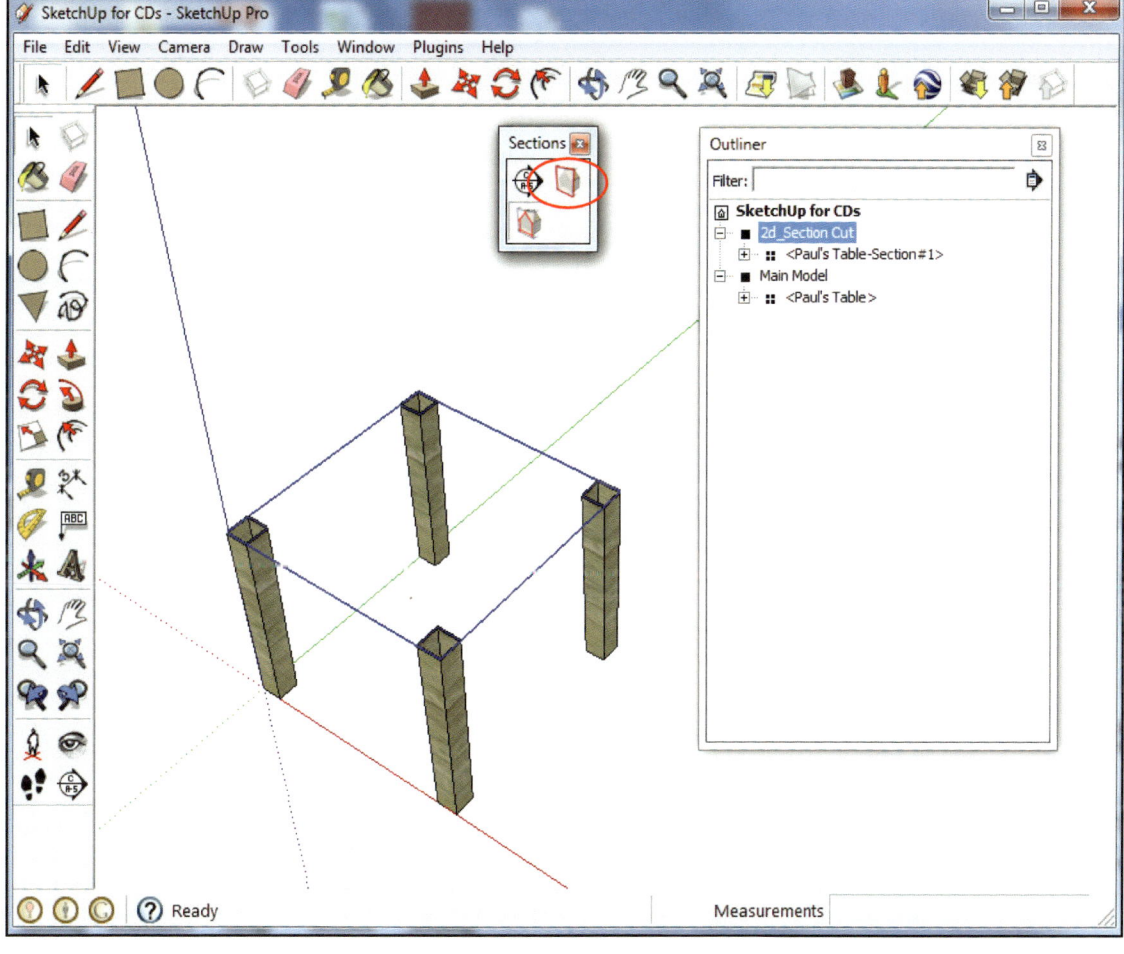

Open the 2D Section Cut Group

✓ Right-click and select "Edit Group"

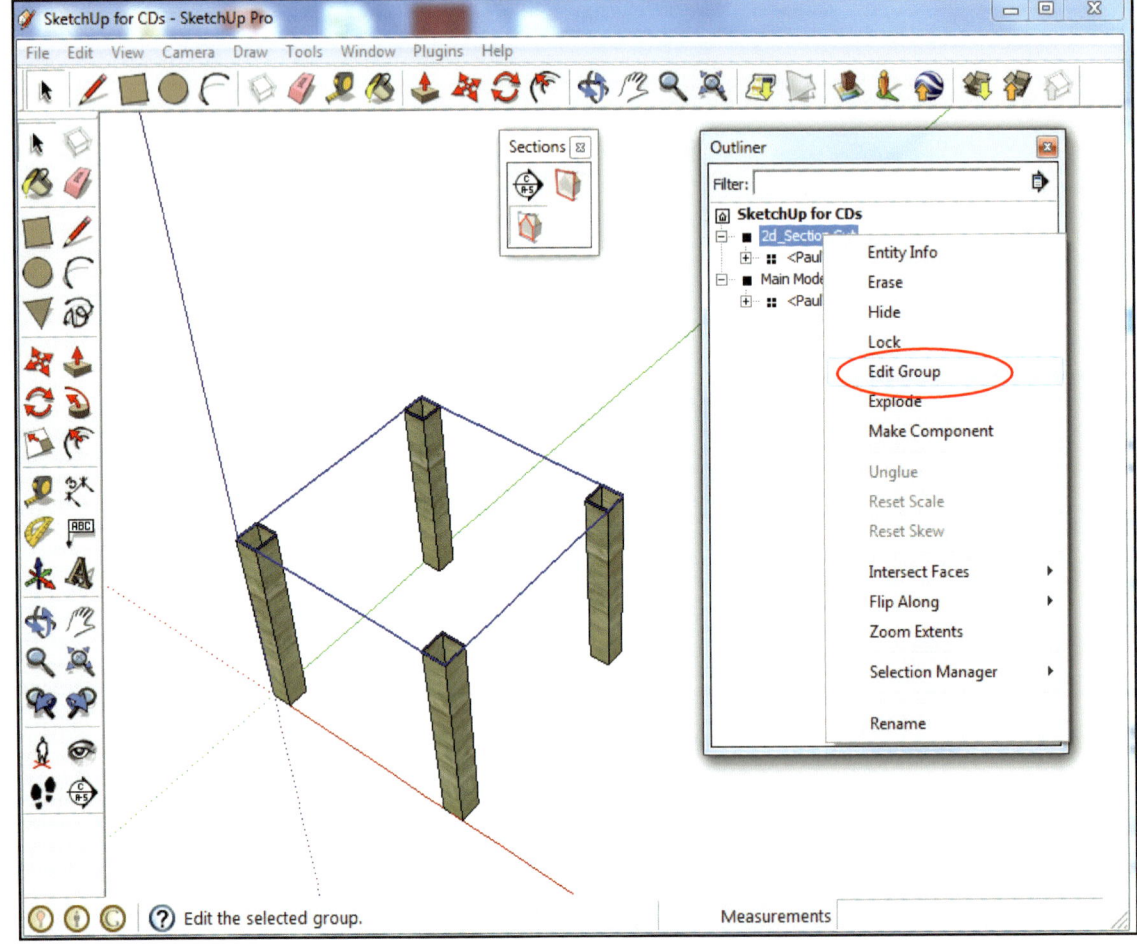

Expand the contents of the 2D Section Cut Group

✓ In Outliner, click on "+" to expand the contents of the Group

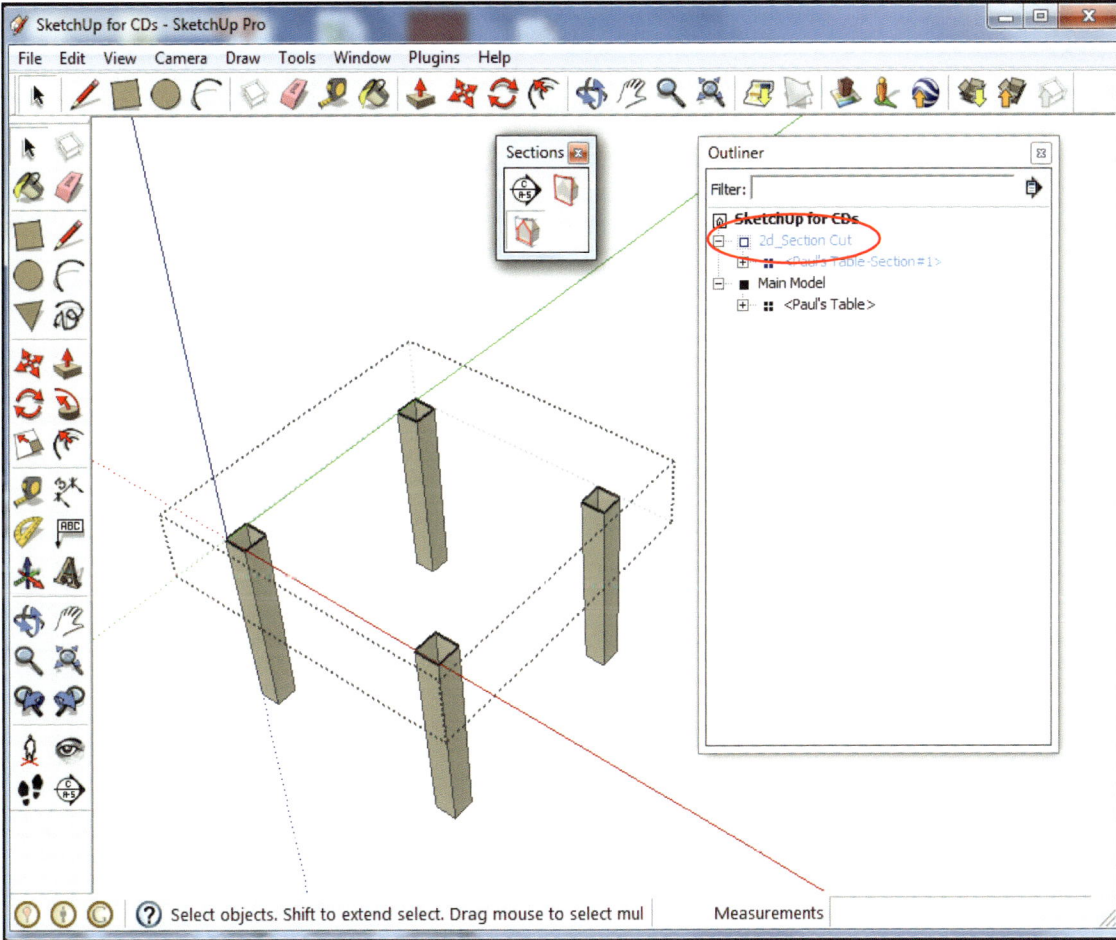

"Hide Rest Of Model"

✓ Activate "Hide Rest Of Model" to visually isolate the 2D Section Cut.

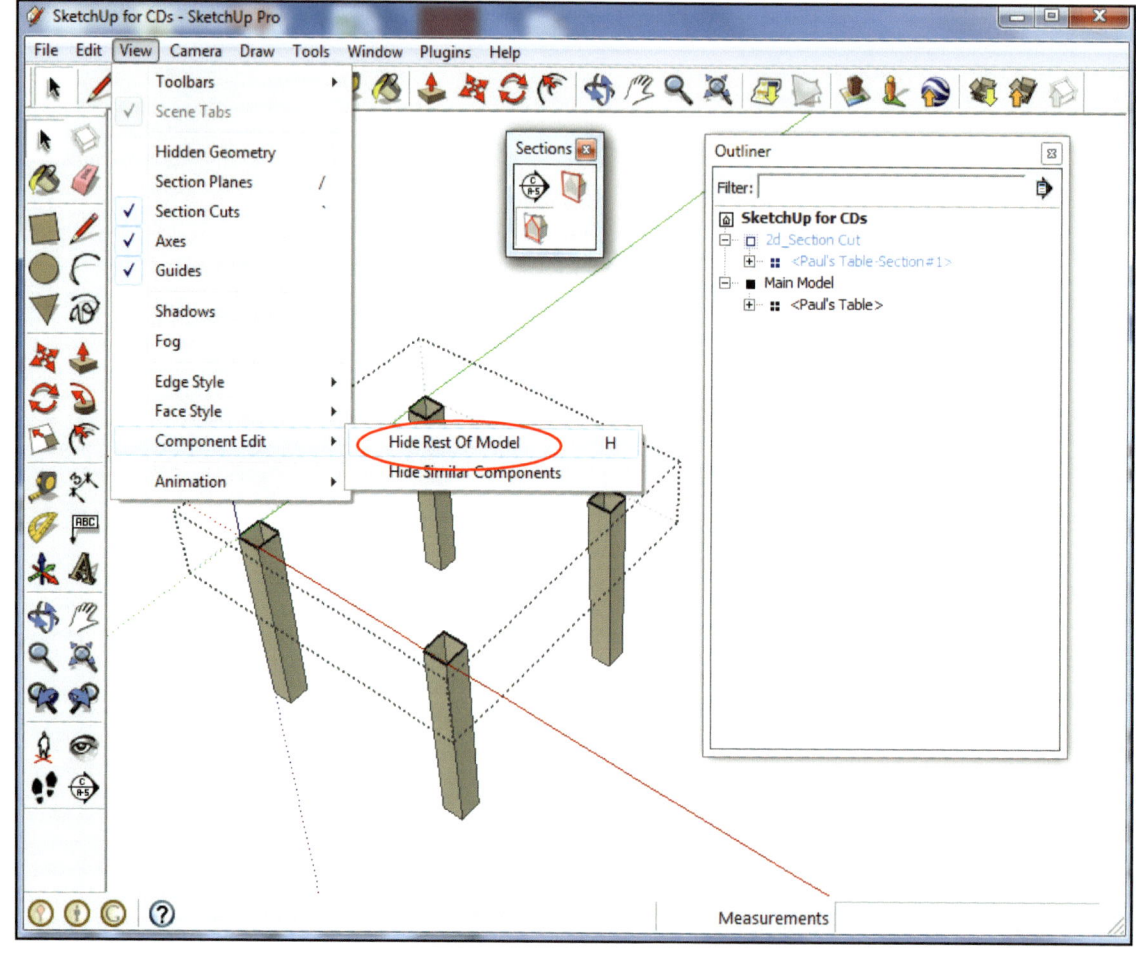

Rest of Model Now Hidden

Explode Contents of 2D Section Cut

✓ Right-click on the 2D Section Cut and select "Explode"

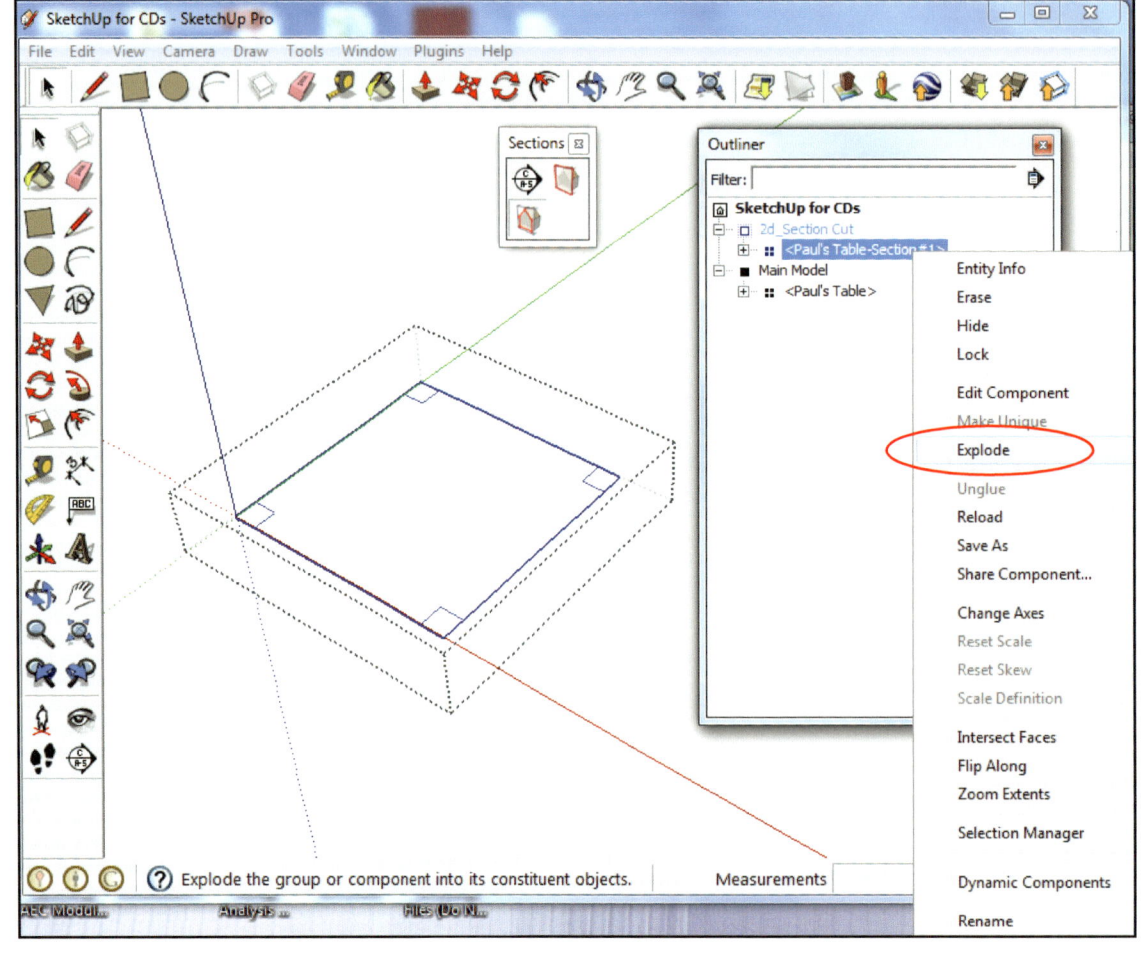

Edit One of the Sub-Components

- ✓ Explode the contents of the 2D Section Cut until there are only the Sub-Components left.
- ✓ (Note: Components are denoted by an icon that is made of four squares, while Groups are denoted by an icon made of just a single square.)
- ✓ Right-click on the Subcomponent and select "Edit Component"

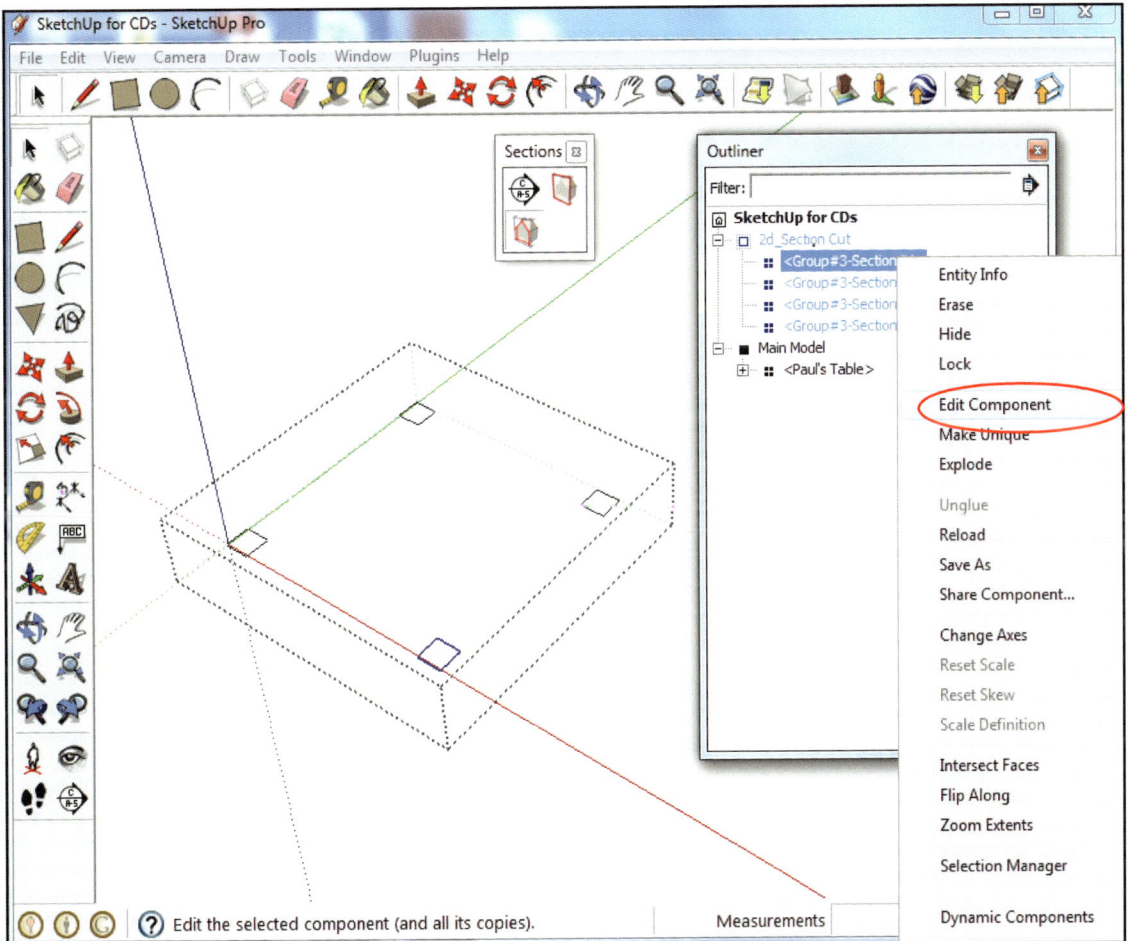

ALWAYS REMEMBER: A Group/ Component MUST BE OPEN for you to edit it! Notice how the bounding box (dashed black line) appears because the Group is open

Finish Editing the Sub-Component

- ✓ Redraw one of the edges of the Sub-Component to establish a surface (1)
- ✓ Right-click on the background (That is: outside of bounding box) to call up "Close Component" dialogue
- ✓ Select "Close Component" (2)

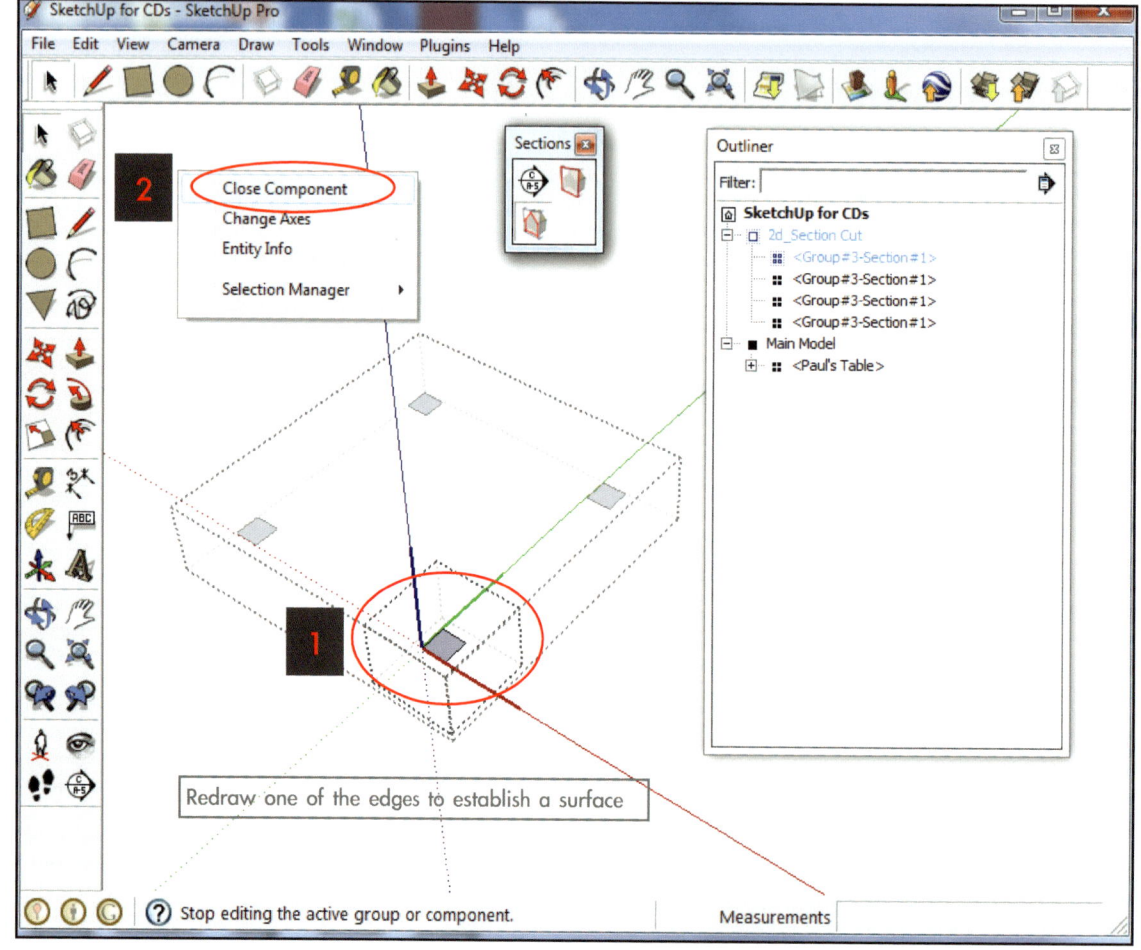

SketchUp2BIM™ Series — Construction Documents in Trimble SketchUp™ Pro by Paul Lee- *A methodology for creating technical 2D drawings*- ©Paul Lee, 2013- www.viewsion.ie

2D Section Cut Closed

✓ Notice that the basic contents of the 2D Section Cut have now been formed, and they are totally independent of the Main Model.

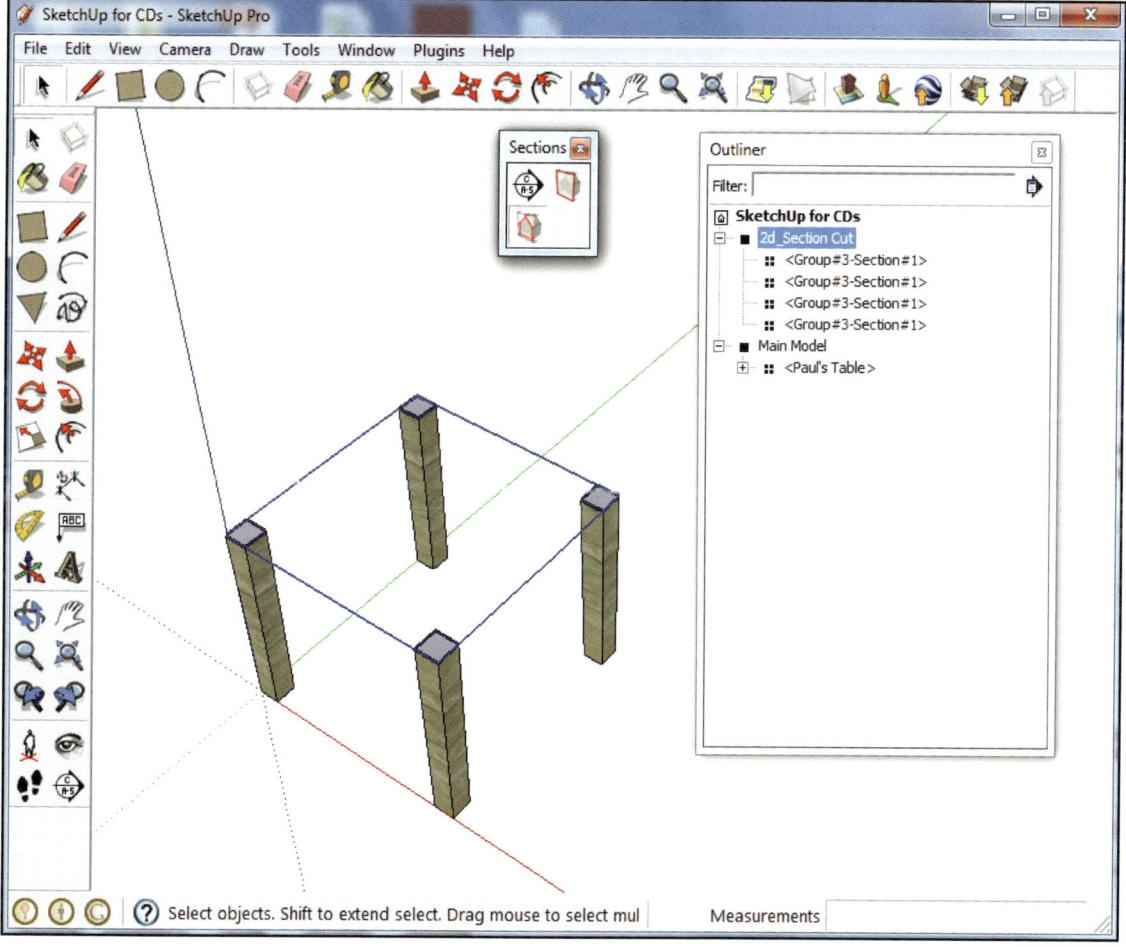

Move the 2D Section Cut away

✓ The Section Cut should hover above the model to avoid clashing with it.
✓ Normally 1" (25mm) or less away from the model.

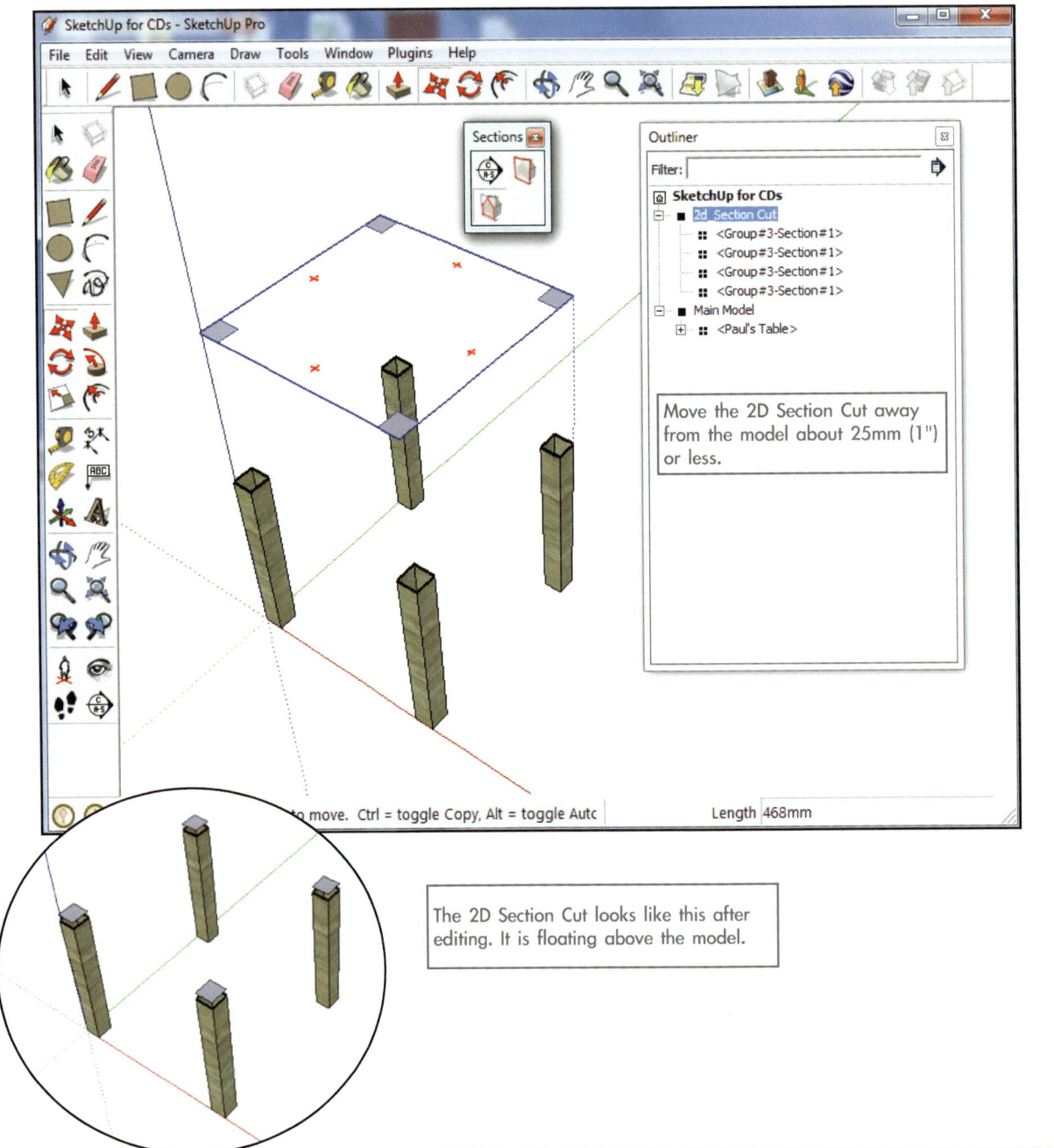

Styles

We will use Styles to great effect in order to emphasize the parts of the model that are in section in contrast to those that are in elevation. Study the diagram below to see the various characteristics of an effective style for viewing an object in section.

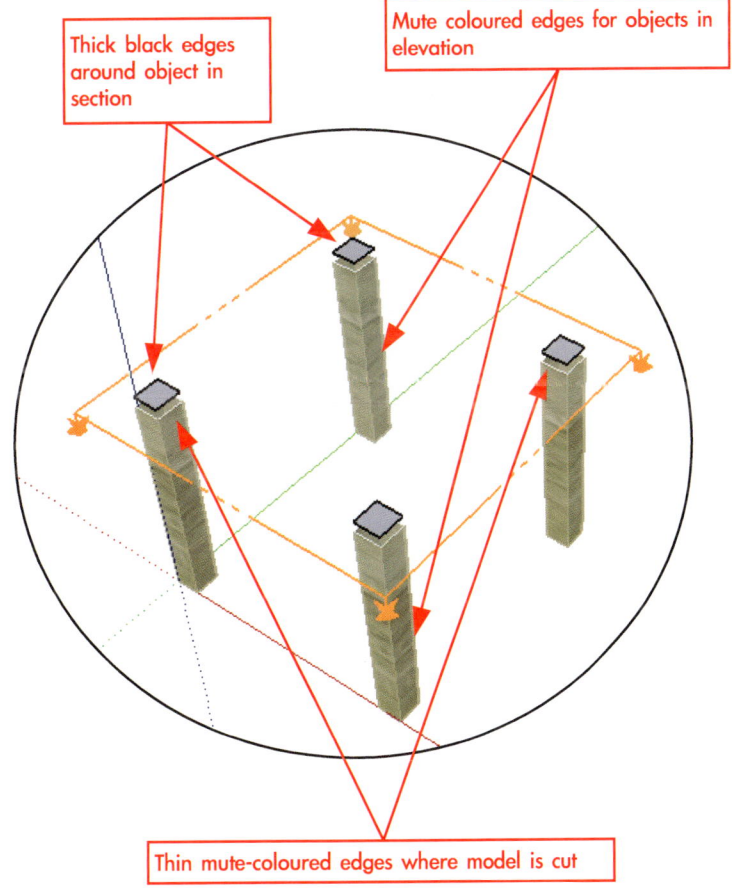

Thick black edges around object in section

Mute coloured edges for objects in elevation

Thin mute-coloured edges where model is cut

Styles Dialogue (1)

- ✓ Go to Window> Styles
- ✓ Initiate a new Style by clicking on the Plus Button (1)
- ✓ Click on the "Edit" tab (2)
- ✓ Select the "Edges" button (3)
- ✓ Turn off all settings except "Edges" (4)
- ✓ Under the "Colour" pull-down menu, select "By Material" (5)

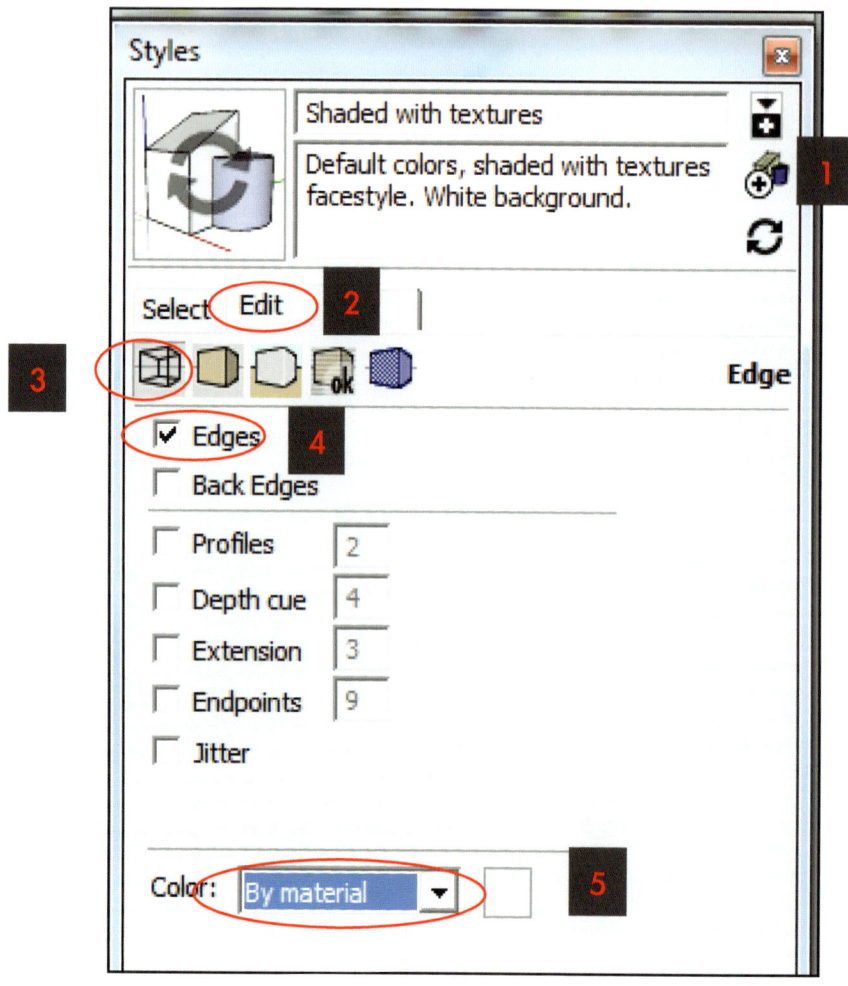

Styles Dialogue (2)

- ✓ Under the "Edit" tab (1)
- ✓ Select the "Modeling" button (2)
- ✓ Click on the "Section Cuts" colour selector and pick a mute colour (3)
- ✓ Under the "Section cut width" change the number to "1" (4)
- ✓ Change the name of the style to "Plan" (5)
- ✓ Click on the icon with the circular arrows to save the settings for this style (6)

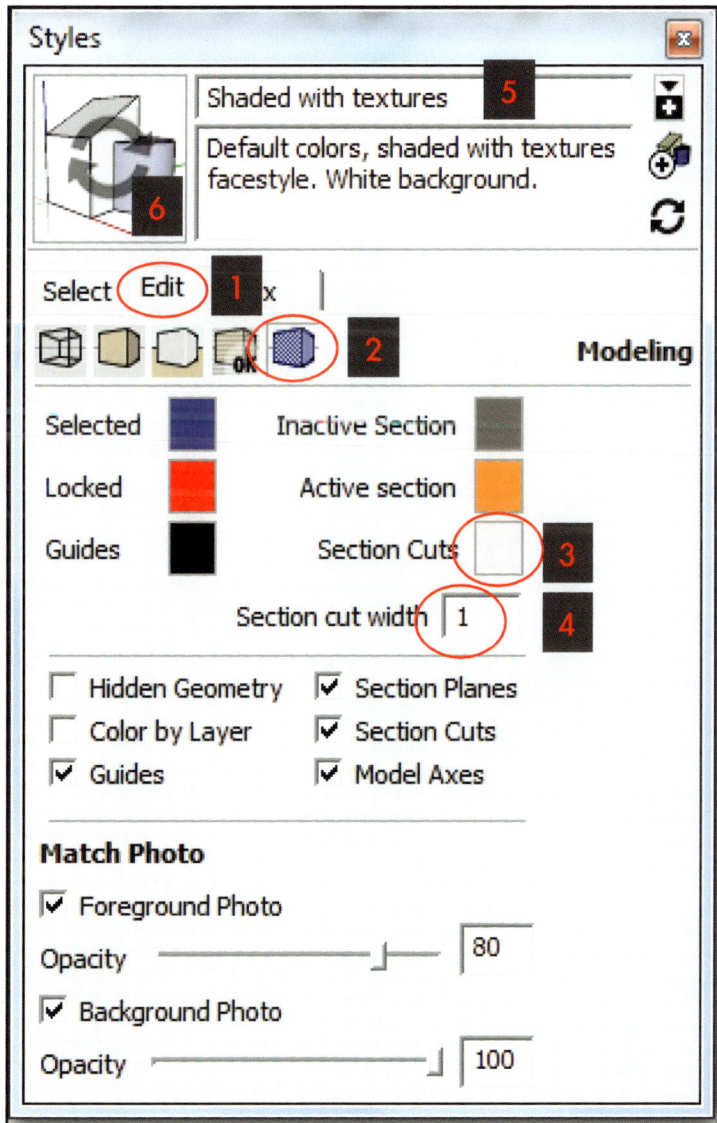

Scenes (1)

✓ To preserve the view of the cutaway model displaying the 2D Section Cut, use the Scenes dialogue to save all of the settings including camera angle
✓ Go to Window> Scenes to bring up the Scenes dialogue

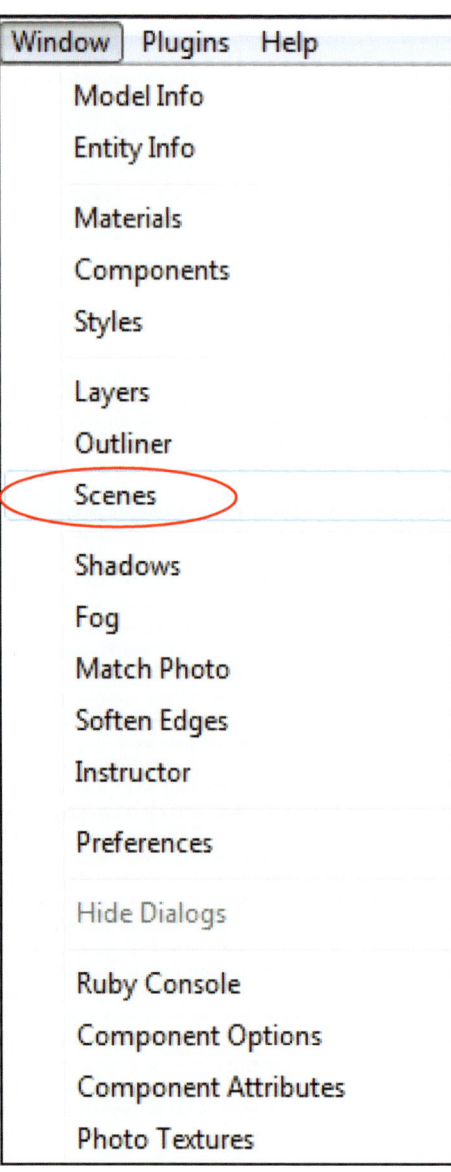

Scenes (2)

✓ To preserve the view of the cutaway model displaying the 2D Section Cut, use the Scenes dialogue to save all of the settings including camera angle
✓ Notice how Styles are a subcomponent of Scenes, so that whatever settings are saved in a particular Style will appear in that Scene (assuming that the "Style and Fog" box is ticked.)

Result

✓ The model should appear like that below, having saved the settings in a new Scene.

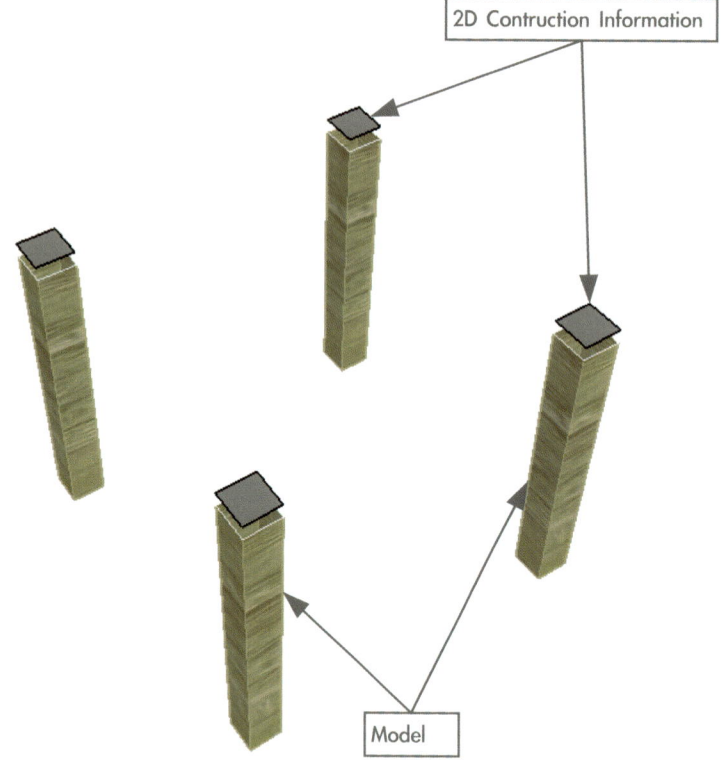

Next: Part 2-
Setting up the Scenes for Importing into LayOut

Creating Construction Documents in Trimble SketchUp Pro

This section takes about 60 minutes to complete

PART 2 of 4

Here I show you how to create the scenes that will be put together to make your construction documents.

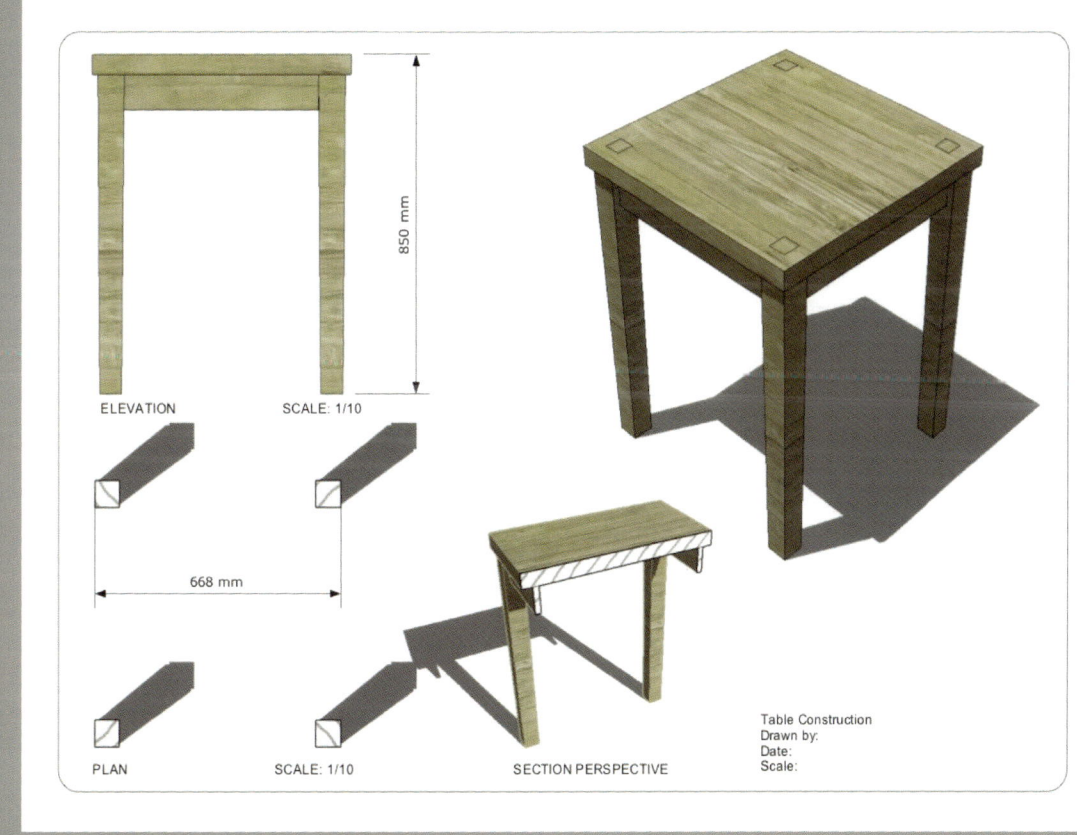

Target Drawing for this Course Series

Creating Construction Documents in Trimble SketchUp Pro

Course Outline

The course is divided into four parts:

Part 1   Creating the 2D construction information.
Part 2   Setting up the Scenes and Importing into LayOut.
Part 3   LayOut: Creating the construction information and Exporting to PDF.
Part 4   Structuring your Model and Recap Exercise.

Part 2

Having used SketchUp to create the plans and sections of your model, in part 2 I show you how to format them into Orthographic (That is: viewed at right-angle & non-perspective) Scenes which are then transferred into LayOut.

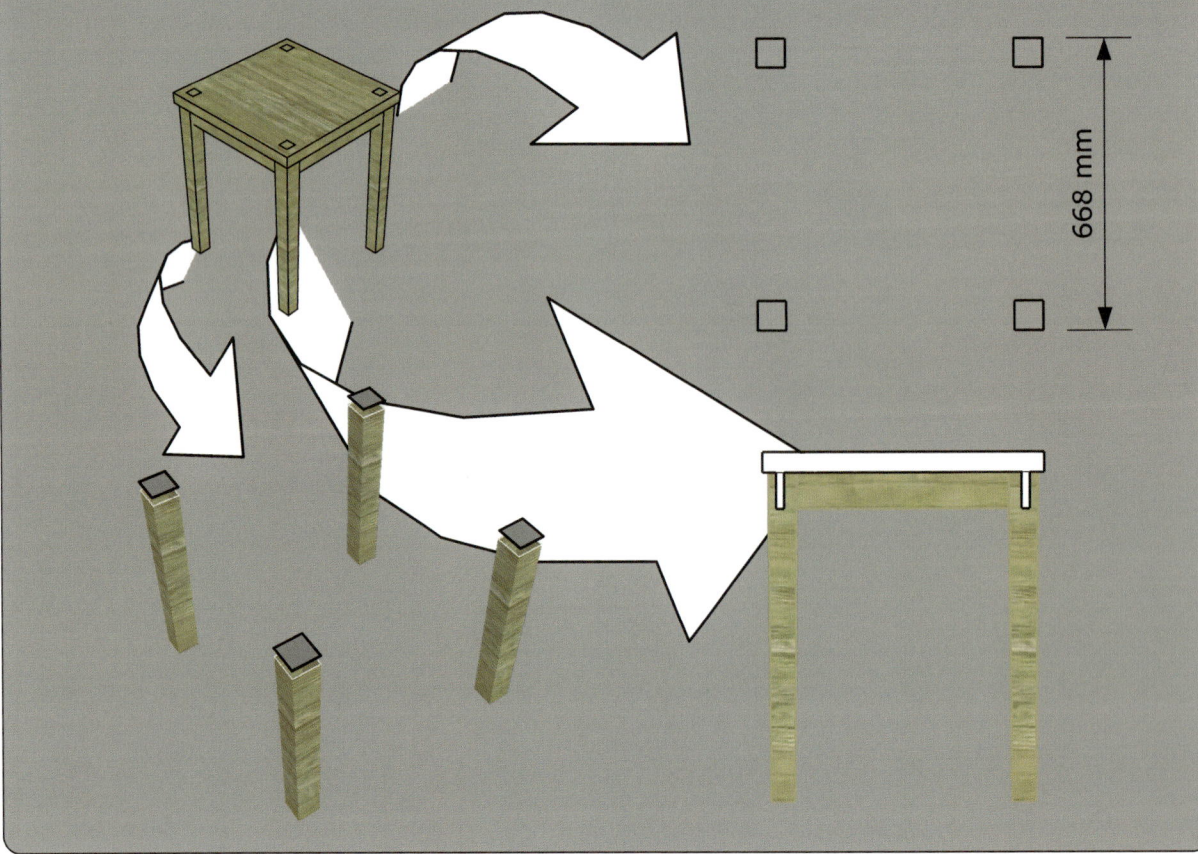

Setting up the Plan (1)

- ✓ Open the SketchUp file prepared in Part 1.
- ✓ Orbit the model to a 45 angle
- ✓ Open the "Sections" Toolbar: View>Toolbars>Sections
- ✓ Make the Plan Section Cut visible (1)
- ✓ Make the same Section Cut active (2)
- ✓ Right Click on the section cut and select "Align View"

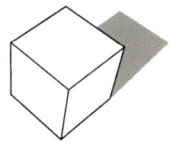

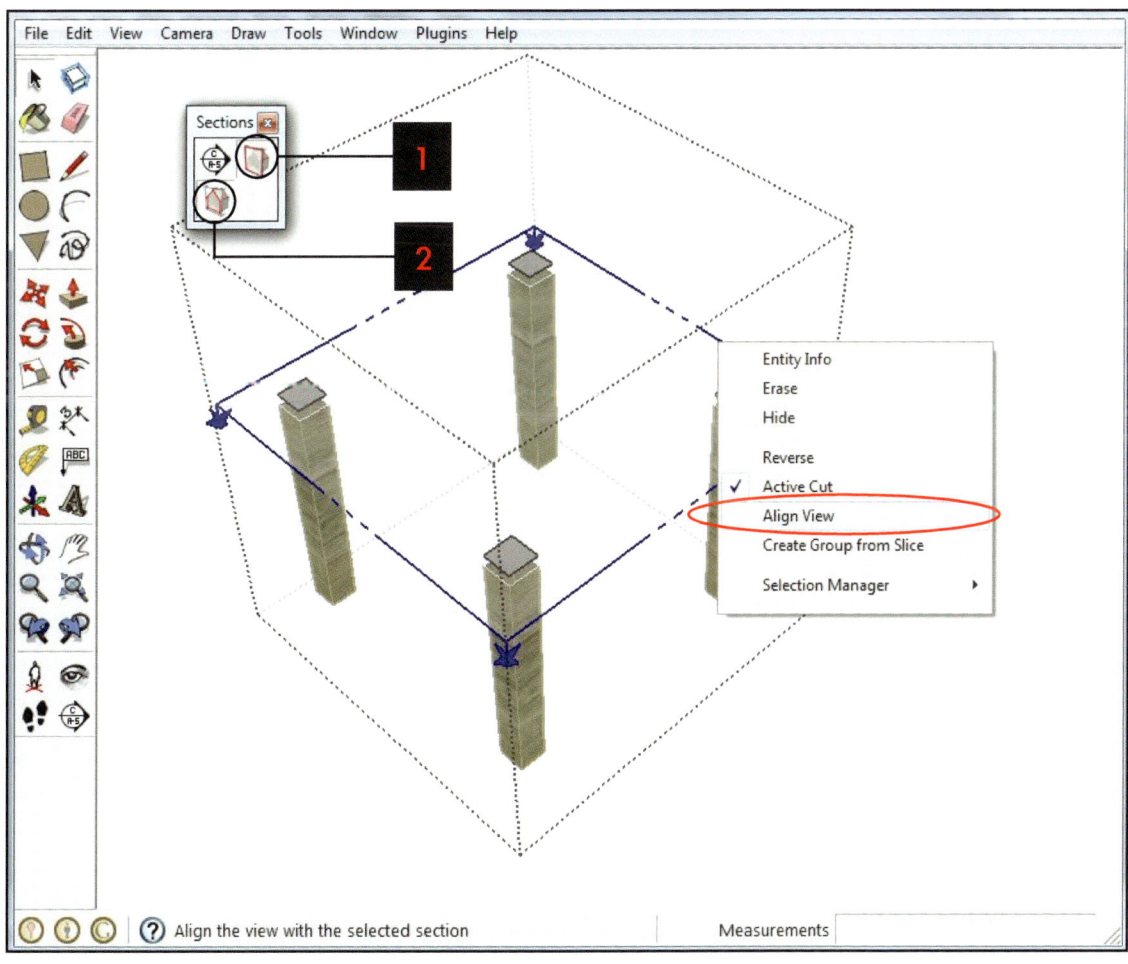

Setting up the Plan (2)

✓ Clicking on the Camera Menu, switch from "Perspective" to "Parallel Projection" (Note: These are mutually exclusive settings- Here I have set up the letter "K" as a shortcut for toggling between the two settings.)

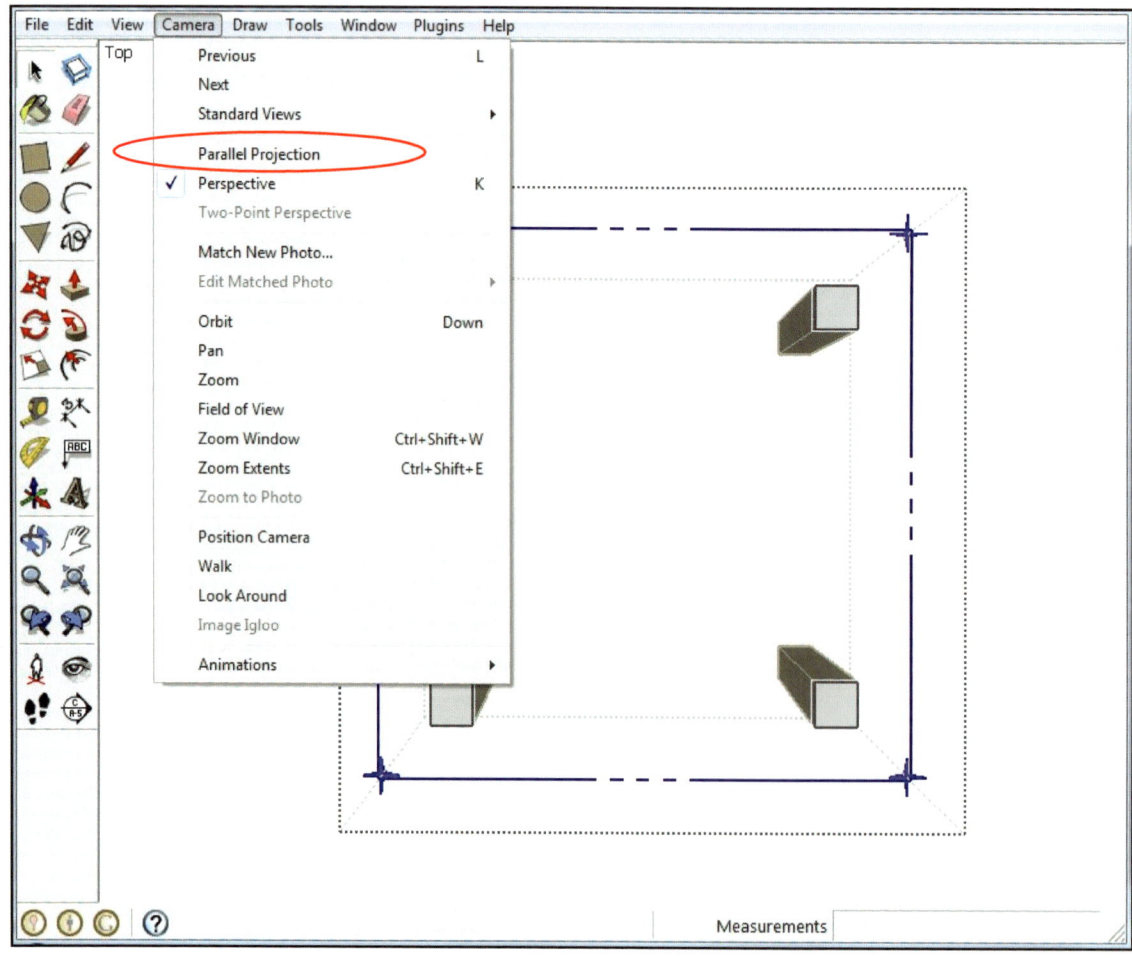

Setting up the Plan (3)

✓ The True Plan View appears

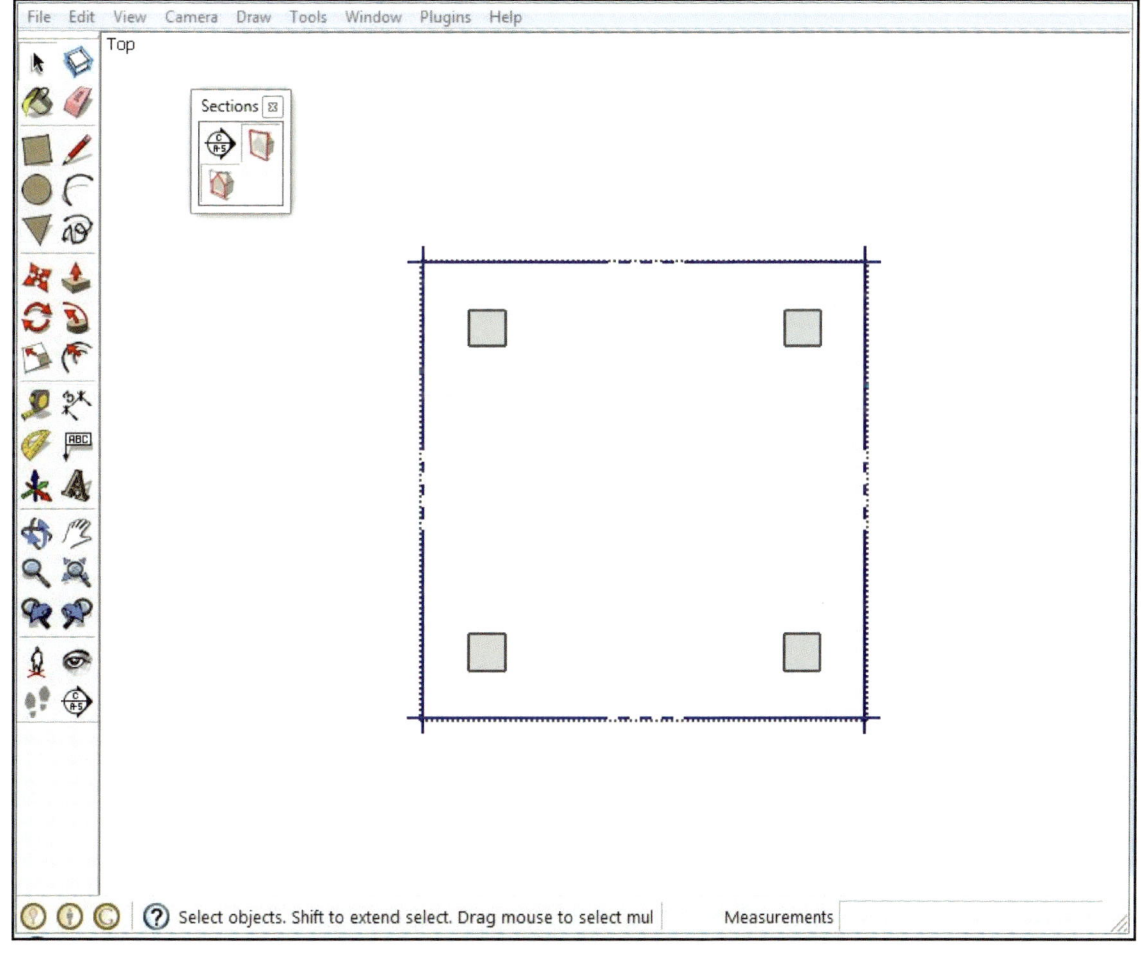

Plan Scene (1)

✓ Open the Scenes Window as illustrated below.

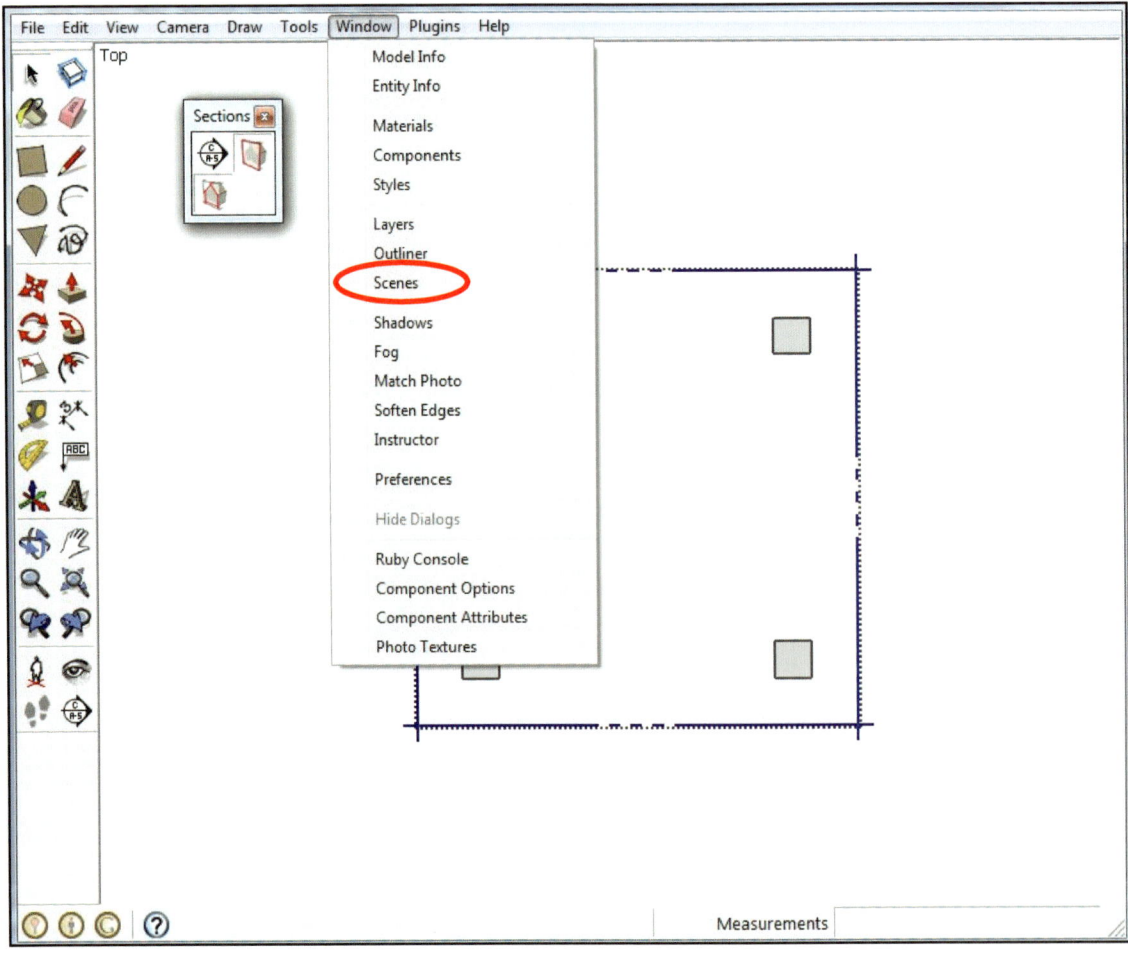

Plan Scene (2)

- ✓ Click on the Plus button to add a new Scene
- ✓ To ensure that the Scenes dialogue is expanded, click on (1)
- ✓ Note the characteristics being added to the scene- those that are selected, such as "Include in animation", "Camera Location", etc.
- ✓ Note here that "Style and Fog" is selected- Style is a fundamental quality to include in the creation of Scenes.

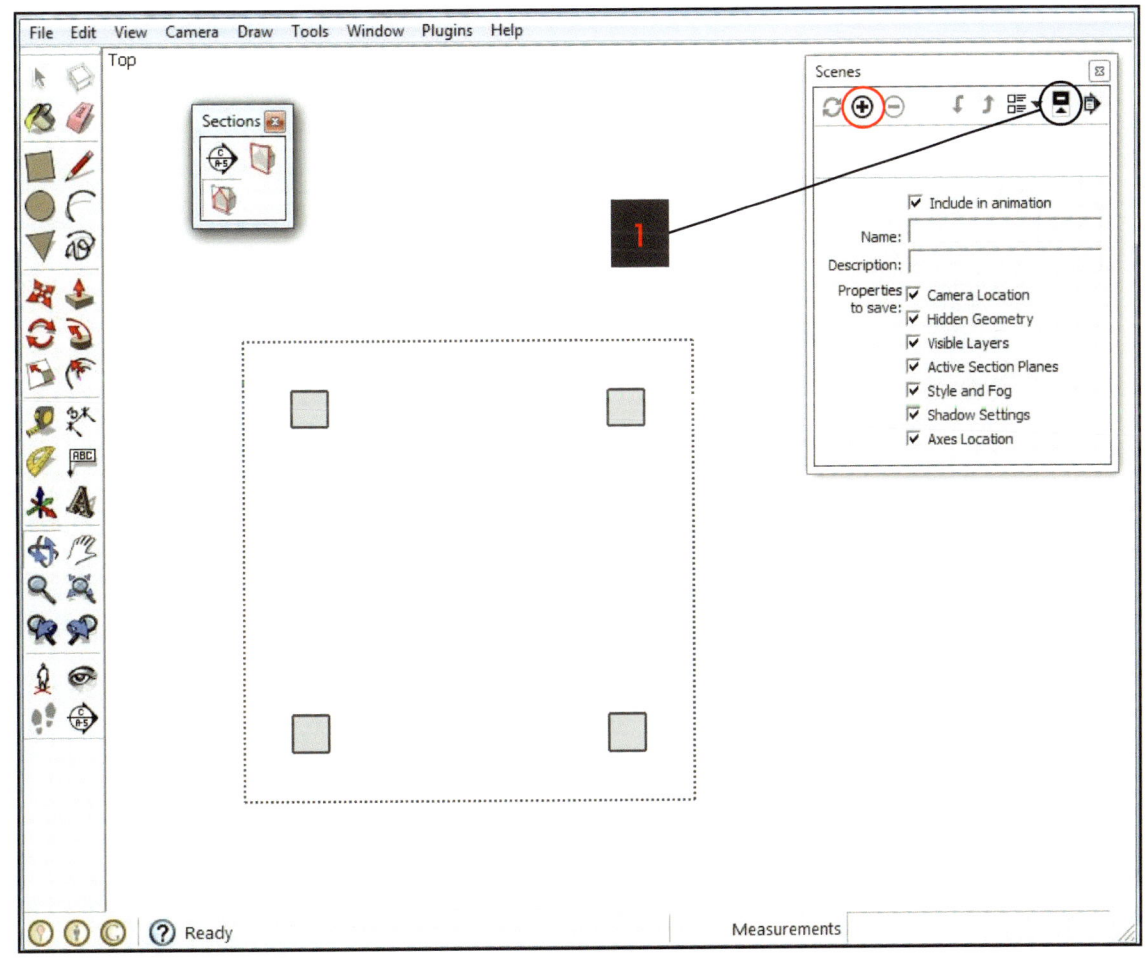

Plan Scene (3)

- ✓ Click into the name dialogue and rename the Scene to "Plan".
- ✓ Orbit to an oblique angle and create a Scene called "Plan Perspective"
- ✓ Now switch off the Section Cut and create a new scene called "Perspective"

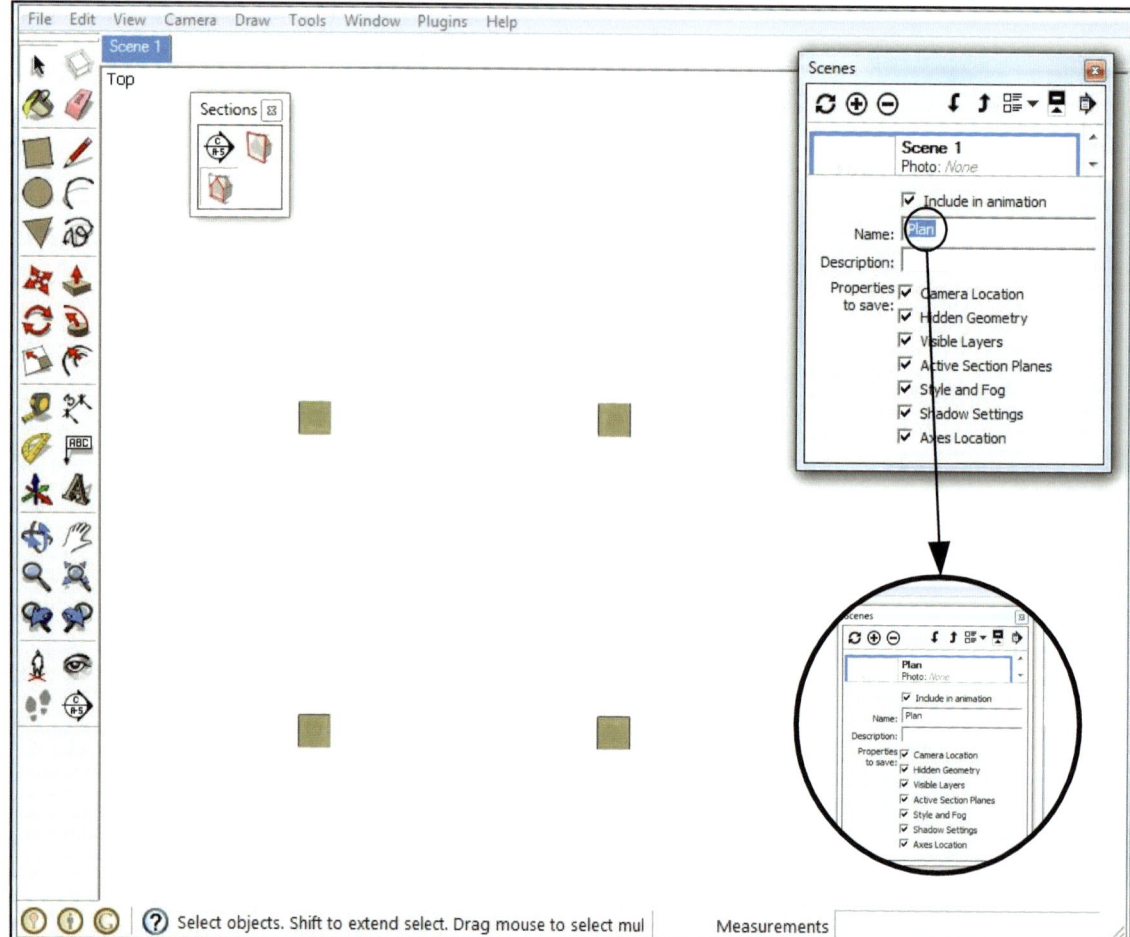

The next section shows how to display the 2D Plan information in the Plan Scene and isolate it from the other Scenes using the "Hidden Layer" plugin,

The Hidden Layer

This section deals with the implementation and use of the "Hidden Layer" plugin.

This plugin allows a layer to be created that is switched off by default in every scene. What is the logic of doing this? Well, when there is an element that you wish to only see in one particular scene (In our example, we only wish to see the 2D Plan information in the "Plan" scene.)

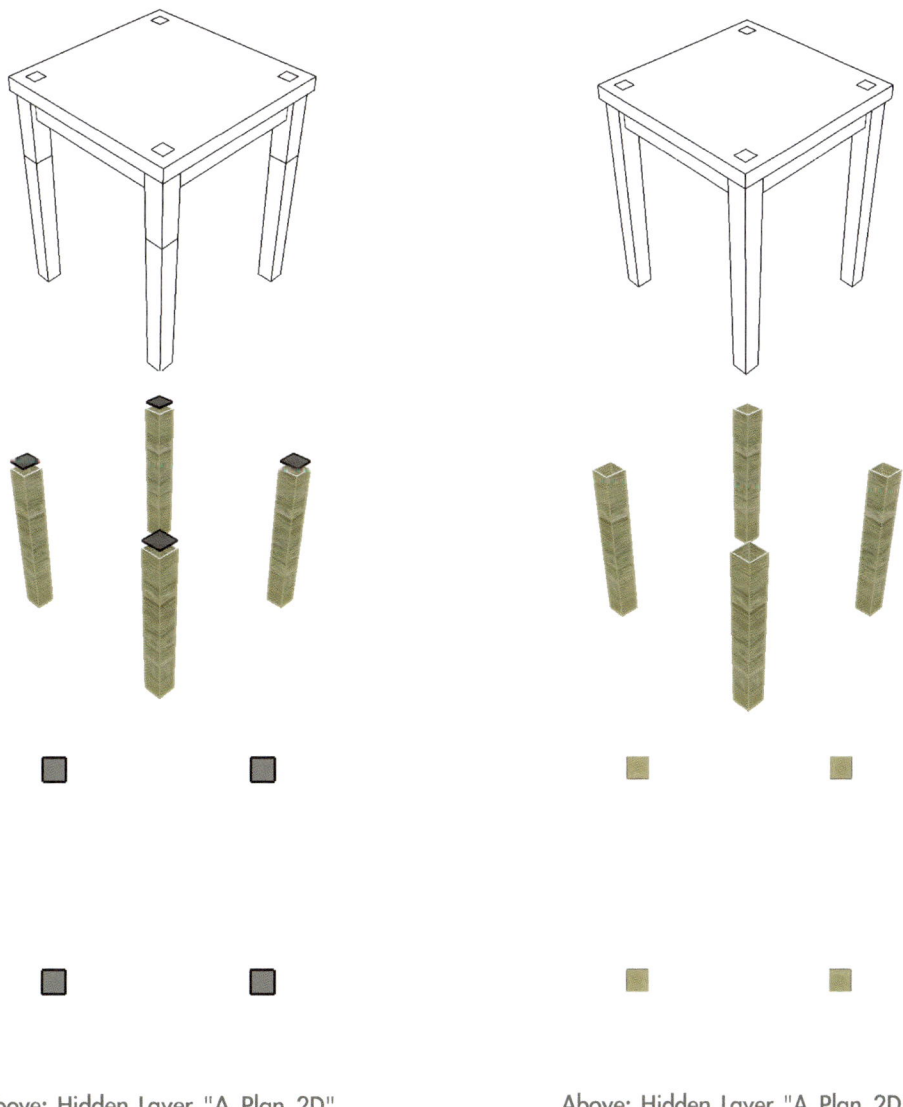

Above: Hidden Layer "A_Plan_2D" switched ON

Above: Hidden Layer "A_Plan_2D" switched OFF

Hidden Layer (1): Implementation

- ✓ Download the "add_hidden_layer.rb" plugin from http://sketchuptips.blogspot.ie/2007/08/add-hidden-layer.html
- ✓ Install the plugin by placing the downloaded file in the correct folder. For information on how to do this: http://sketchuptips.blogspot.ie/2008/03/how-to-download-and-install-plugins.html
- ✓ Unless you haven't done so already, close down and re-open SketchUp so the plugin becomes operational.
- ✓ Using the "Add Hidden Layer" command from the Plugins Menu as below, create two hidden layers called "A_Plan_2D" and "A_Section_A"
- ✓ Note: Hidden Layers are switched off by default in every Scene. This is extremely useful here for controlling the visibility of the 2D section (or plan) information, where the information needs to only appear in one or two scenes.

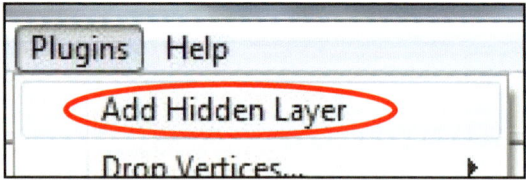

Hidden Layer (2)

✓ Using the Outliner Window to view the model structure, right-click on the 2D Section Cut and select Entity Info

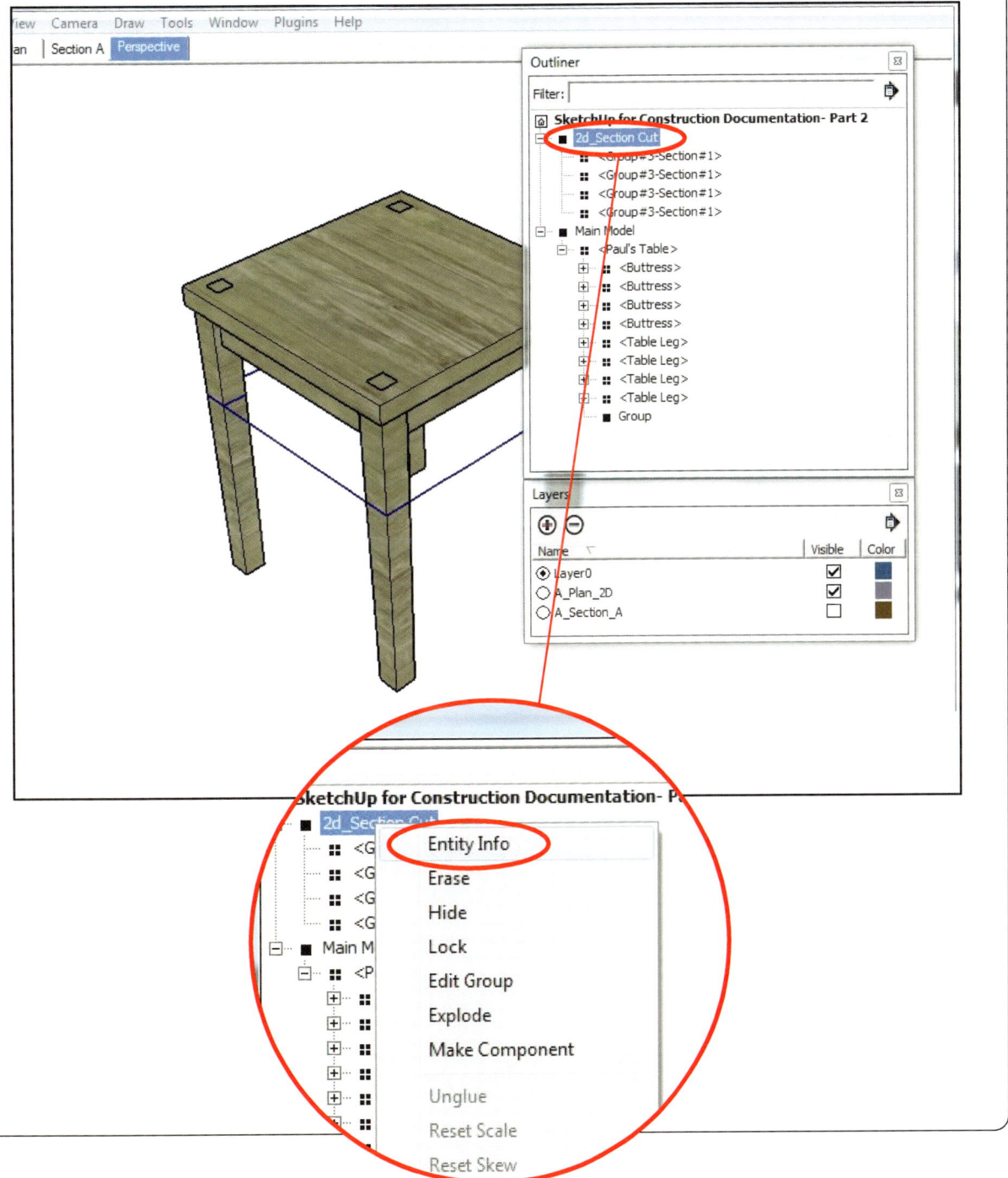

Hidden Layer (3)

✓ Using the Entity Info dialogue, click on the Color button (1) and change it to Black (This means the 2D information will show up black when lines are set to colour "by material"- see Styles in Part 1.)
✓ Next, under the Layer pull-down menu, select the "A_Plan_2D" layer (2).
✓ Note: Transferring the 2D Section Cut onto a hidden layer makes the entire entity disappear when a Scene is created. This means it disappears in the Outliner dialogue also.

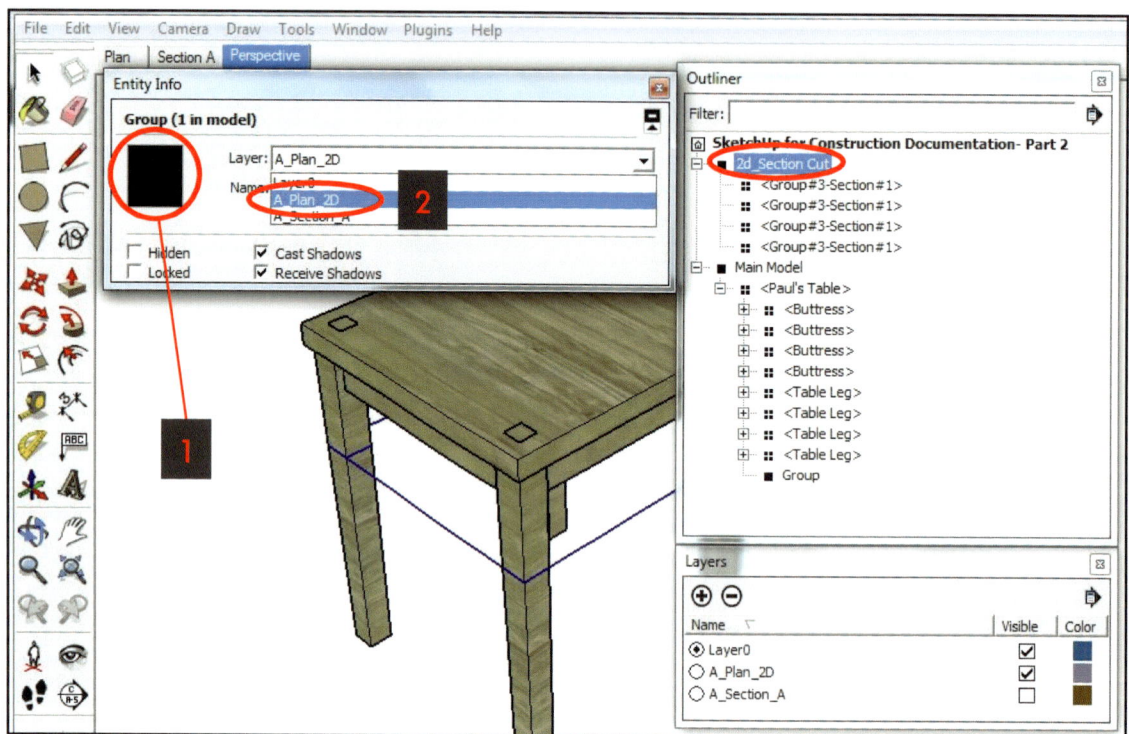

The 2d_Section Cut disappears when:

✓ It has been placed on a hidden layer, and;
✓ A Scene has been created

Turning On the Hidden Layer

✓ Within the Plan Scene, the 2D Section information can only be turned on using the Layer Dialogue.

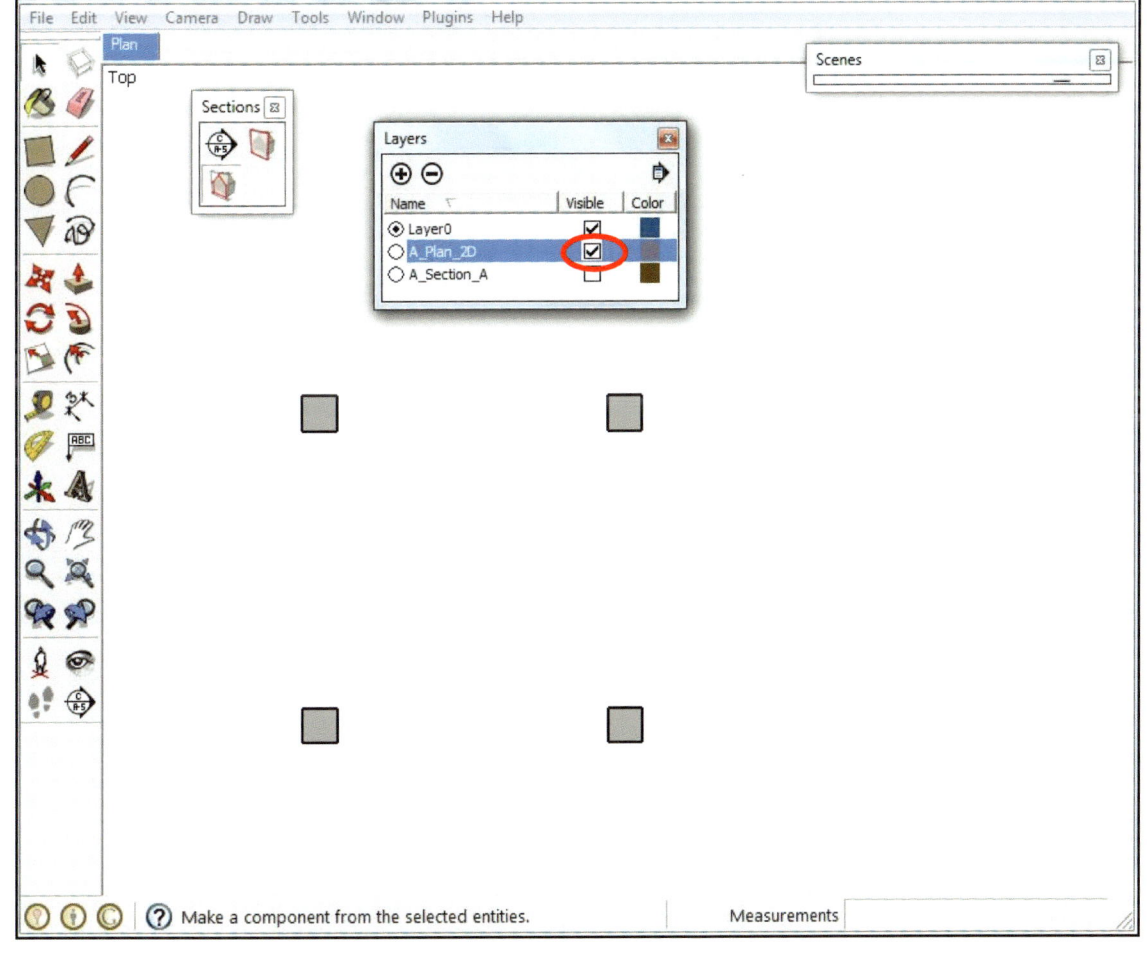

Saving the Scene with Hidden Layer Switched On

✓ Once the layer has been turned on, THE SCENE MUST BE UPDATED by right-clicking on the Scene Tab and selecting "Update" (1), or by clicking on the update icon in the Scenes dialogue. (2)

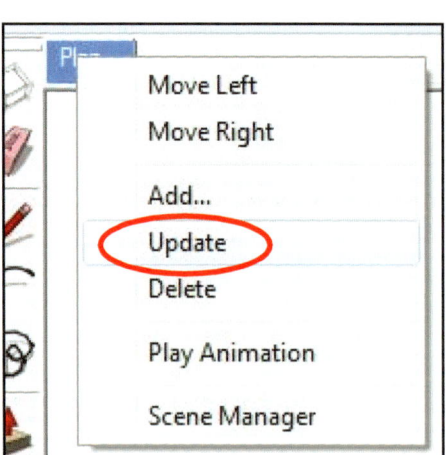

Creating the Vertical Section Cut

✓ To Create The Section Scene, the same steps follow as for the Plan Scene, the difference being that the 2D_Section Cut entity is put on it's own Hidden Layer which is switched on for that Scene.

Note: There are two ways to achieve the view of the Section Cut opposite. One is the previous method of "Align View" to the Section Cut. The other is to use the Standard Views Toolbar (Find this under View>Toolbars>Views) Using these tools, select "Front" view and then ensure that Perspective is turned off (Camera>Parallel Projection)

Result

Now you have created the Scenes you require to show in LayOut.

Next: Part 3-
LayOut: Scene insertion, manipulation and graphic overlays including dimensions.

Creating Construction Documents in Trimble SketchUp Pro

> This section takes about **60** minutes to complete

PART 3 of 4

Here I show you how to insert your Construction Drawing Scenes into LayOut, scale and style them, implement graphics (including dimensions) to finalise your construction documents.

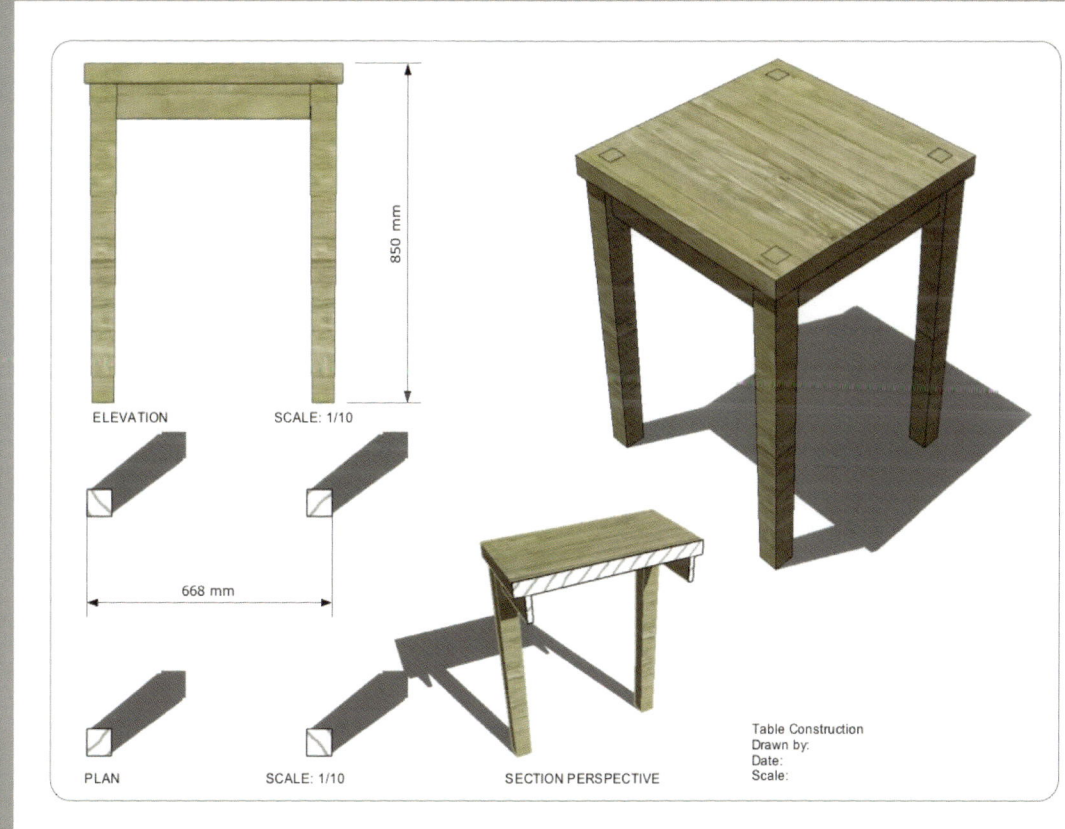

Target Drawing for this Course Series

Using LayOut

- ✓ Open LayOut
- ✓ Under Templates, choose A4 Landscape

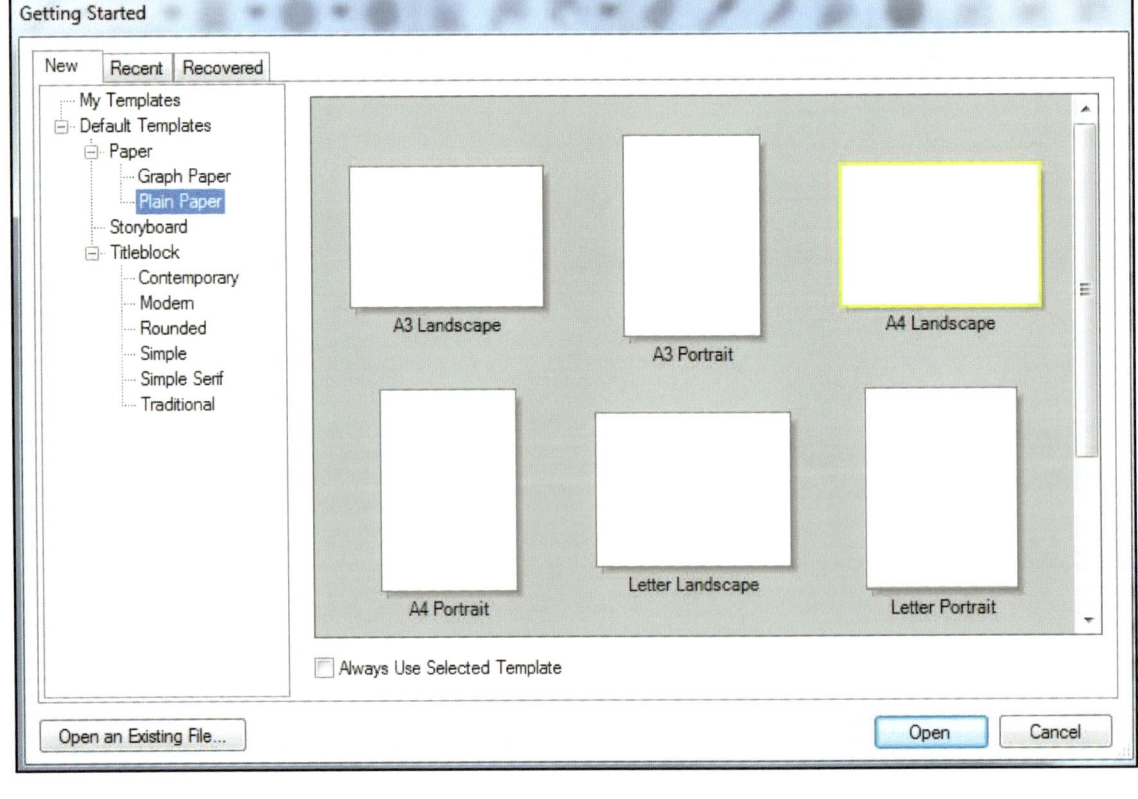

SketchUp2BIM™ Series — Construction Documents in Trimble SketchUp™ Pro by Paul Lee- *A methodology for creating technical 2D drawings*- ©Paul Lee, 2012- www.viewsion.ie

Creating the Titleblock

- ✓ Click on the Rectangle Tool, and hold to get pulldown menu.
- ✓ Select Rounded Rectangle.

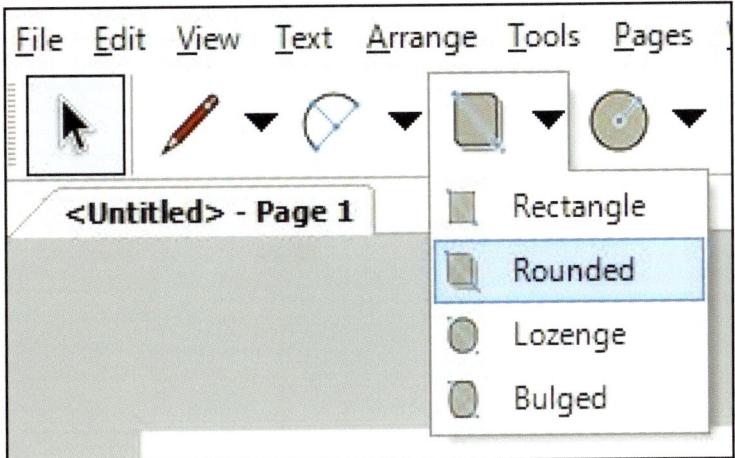

- ✓ Draw a rectangle across the page by dragging from corner to opposite corner- While making the rectangle, you can use the up and down arrow keys to increase/reduce the radius of the corners.

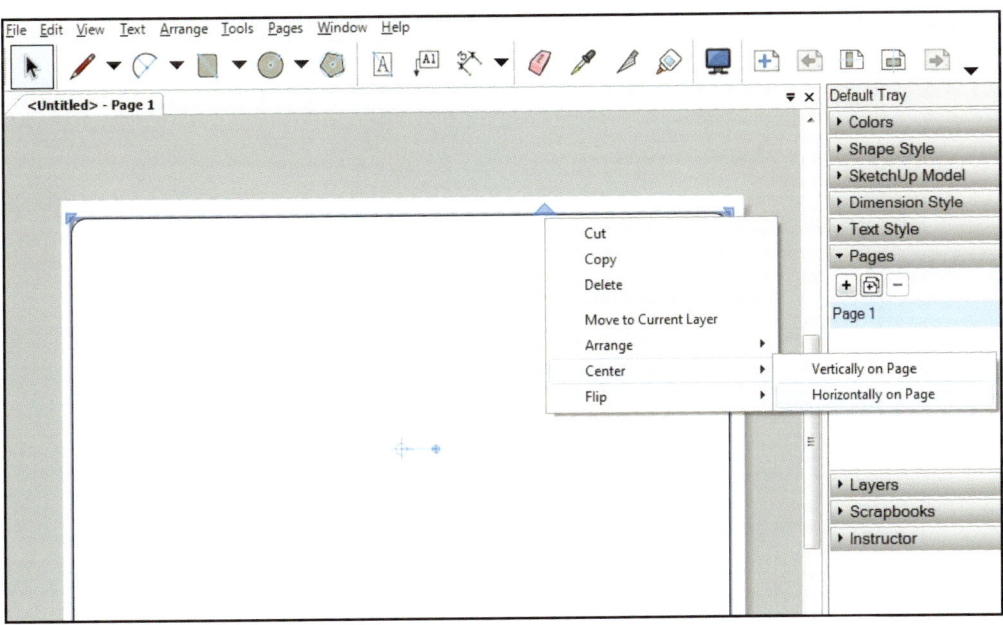

Inserting the Model

- ✓ Go to: File>Insert and select the Model
- ✓ When the model appears, right click on the model itself
- ✓ Select Scenes>Perspective
- ✓ Note: Perspective is one of the Scenes that were set up in the SketchUp model. If you create a new Scene within the Model, you must save it in the SketchUp file and then right-click>Update to see the new Scene and/ or Style.

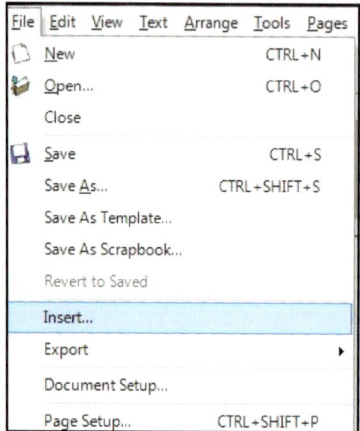

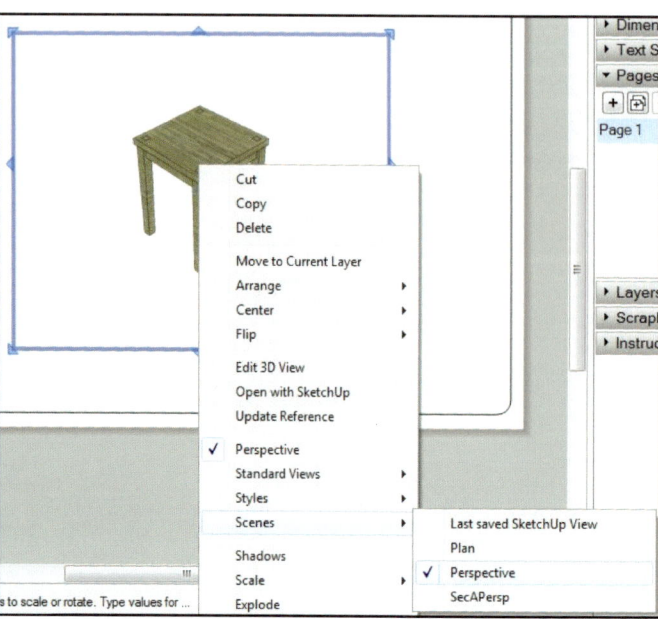

Creating different views of the Model (1)

✓ Copy the model view (Using the Move Tool with Ctrl key pressed.)
✓ Right-Click on the model view and pick another Scene.

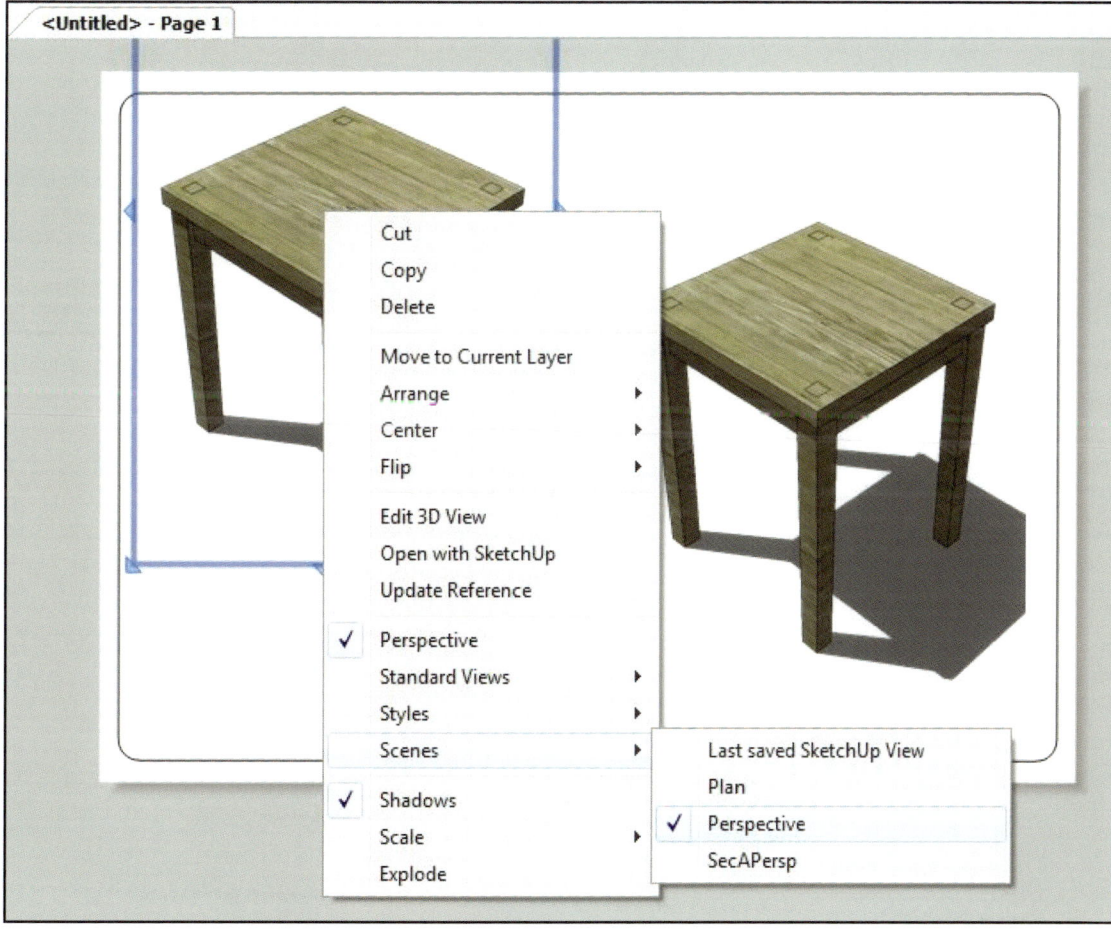

Scaling the View

✓ There are two ways to view properties such as model view, perspective, standard views and scale:

    1   Right-click on the model view to obtain menu options.

or

    2   Use the "SketchUp Model" control panel (Found under the Windows Menu)

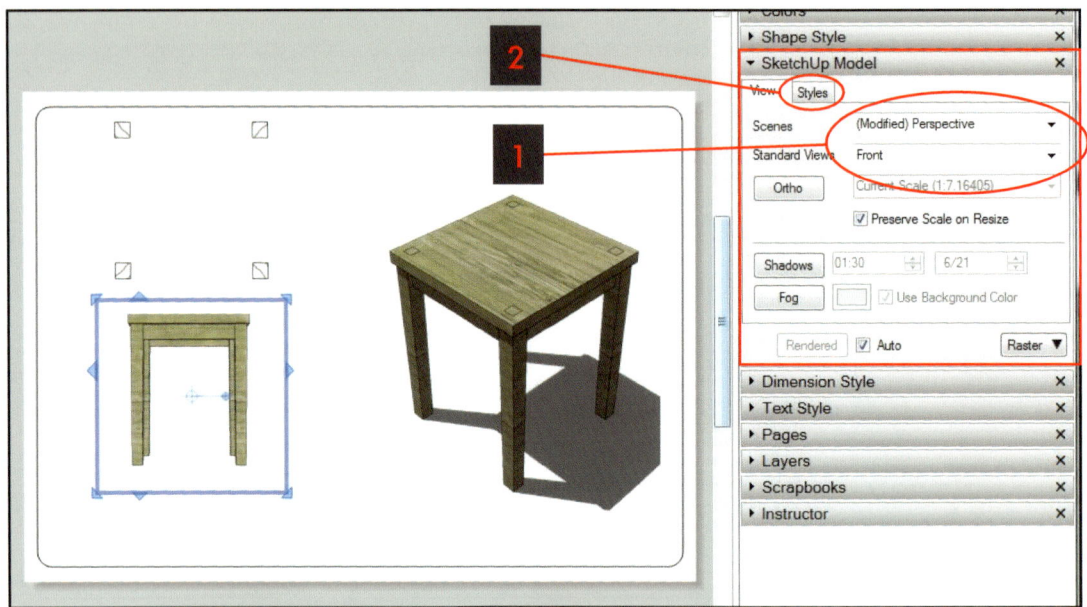

✓ Notice in the control panel: Various options available to use your pre-formed Scene or a Standard View (1). Also available are shadows, fog, preserve scale on resize (for non-technical scaling.)
✓ On the second tab (2) Styles can be edited.
✓ You can also go freestyle with any view by double-clicking into the drawing. This allows you to orbit, zoom and pan within the Model View
✓ When you select a Model View, you can use the grips to resize the Window or the "Lever" (The circular dot in the middle) to rotate the View or create a new reference point for placement.

SketchUp2BIM™ Series   Construction Documents in Trimble SketchUp™ Pro by Paul Lee- *A methodology for creating technical 2D drawings-* ©Paul Lee, 2012- www.viewsion.ie

Inserting Dimensions

- ✓ The Dimensions Tool is available from the top toolbar.
- ✓ Dimensions automatically adjust to the scale of the view.
- ✓ Ensure that Object Snap is switched on: Arrange: Object Snap

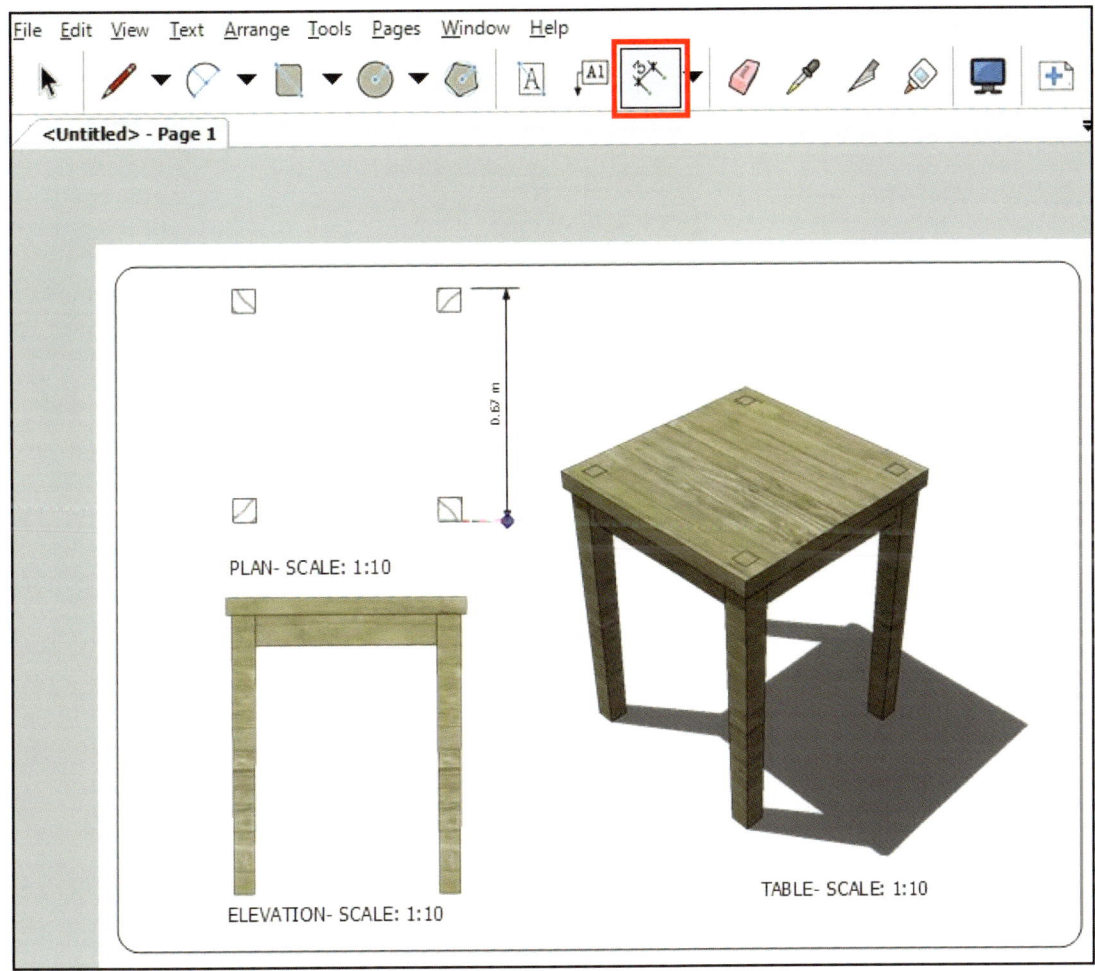

Controlling the Dimension Style

- ✓ Dimension Styles have their own Control Panel.
- ✓ They automatically adjust to the scale of the drawing you are referring to.
- ✓ Explore the various options available, such as alignment, units etc.
- ✓ Note that the styles of the lines and text are controlled using the Shape Style and Text control panels.

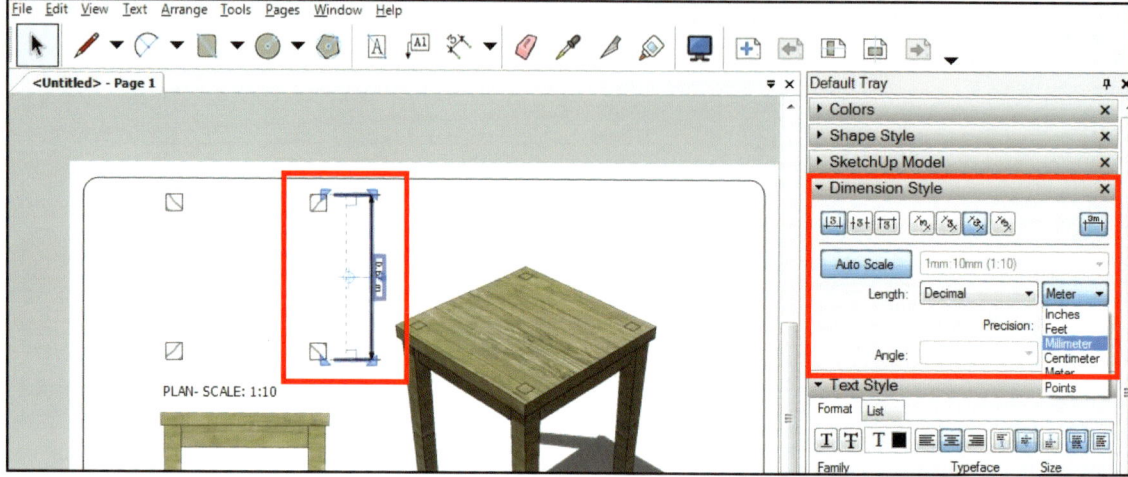

Document Settings

✓ To control document setup for items such as default dimensions, paper size, grid etc., click on Edit>Document Setup and select from the menu on the left.

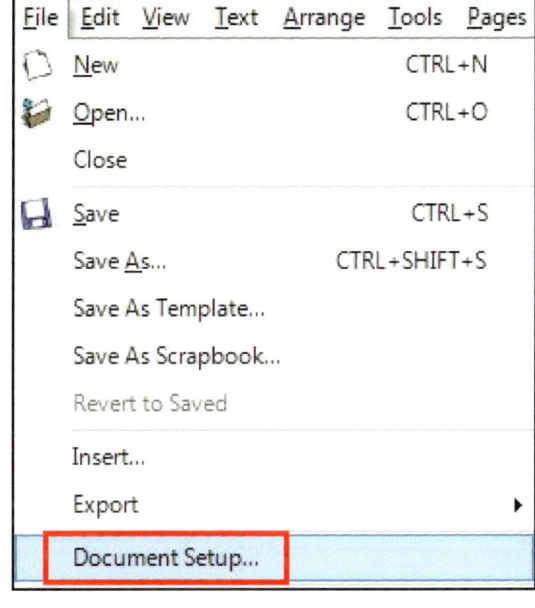

Text Editing

- ✓ Creating Text is extremely easy in LayOut, utilising a simple rectangle layout within which you type your text.
- ✓ To edit text, use the dialogue found under the Windows Menu

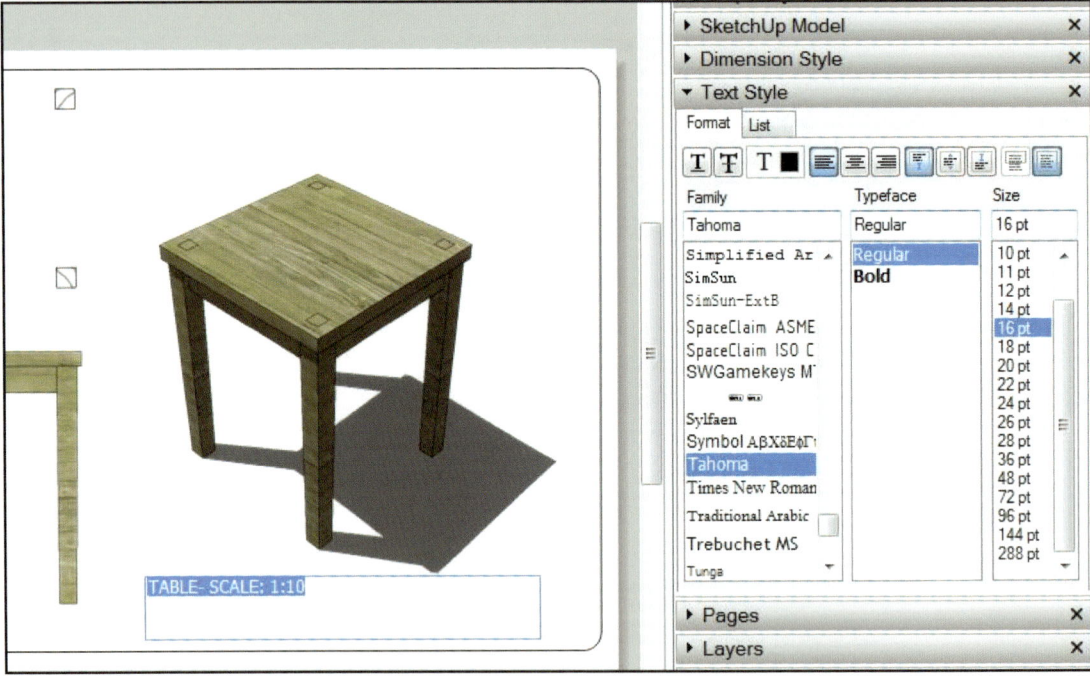

01
02
03
04
05
06
07
08
09
⑩
11
12
13

SketchUp2BIM™ Series Construction Documents in Trimble SketchUp™ Pro by Paul Lee- *A methodology for creating technical 2D drawings*- ©Paul Lee, 2012- www.viewsion.ie

Model View Styles (1)

✓ Vector is a format that allows your SketchUp Model to be viewed like a fully-fledged CAD drawing. All lines are smooth and defined.
✓ The Hybrid setting allows textures to be viewed while applying Raster lines.

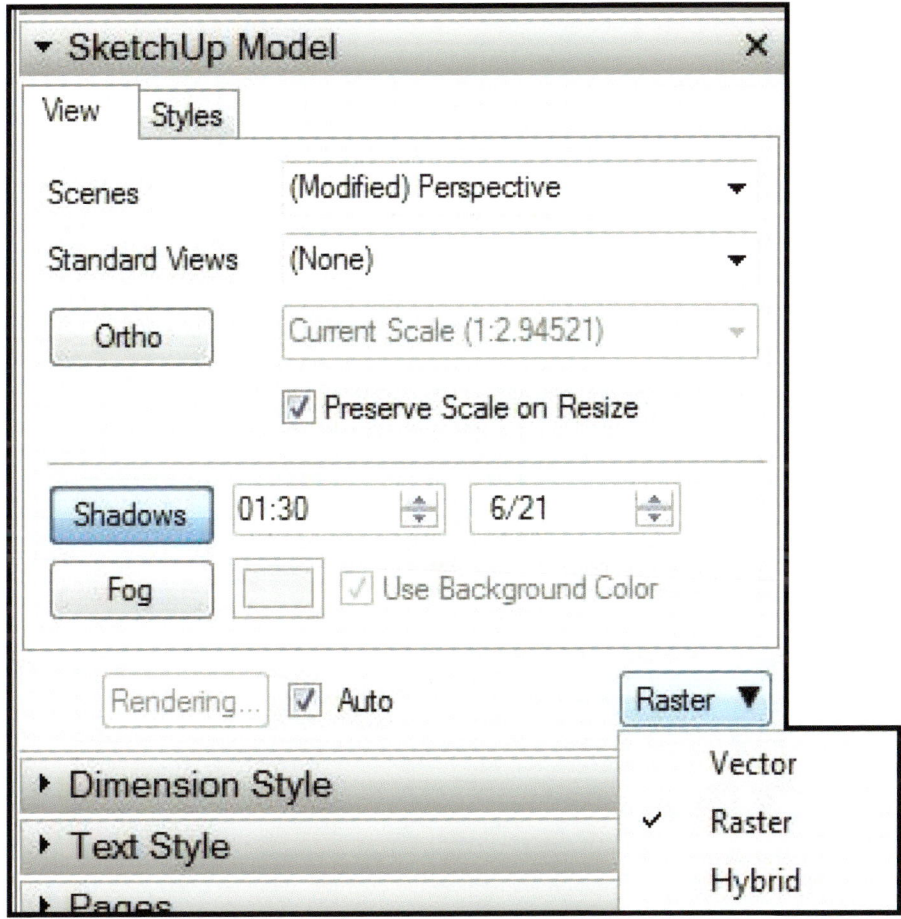

Note: It is better to use the Raster setting while you are preparing your documents. Vector and Hybrid settings are more taxing on your computer and can add considerably to your work time. These more refined settings should generally only be applied after the document has been set out and prior to pdf export/ print.

Model View Styles (2)

- ✓ Under the SketchUp Model menu, the second tab controls Styles (1).
- ✓ In the "Home" view you can see the Styles that were created in your model (2).
- ✓ You can also access the pull-down menus on the right to access standard Styles (3).
- ✓ You can control the lineweight by clicking into the numerical dialogue (4).
- ✓ The raster/ vector/ hybrid are controlled using the pull-down menu (5).
- ✓ You can turn on and off the background by clicking on the checkbox (6).

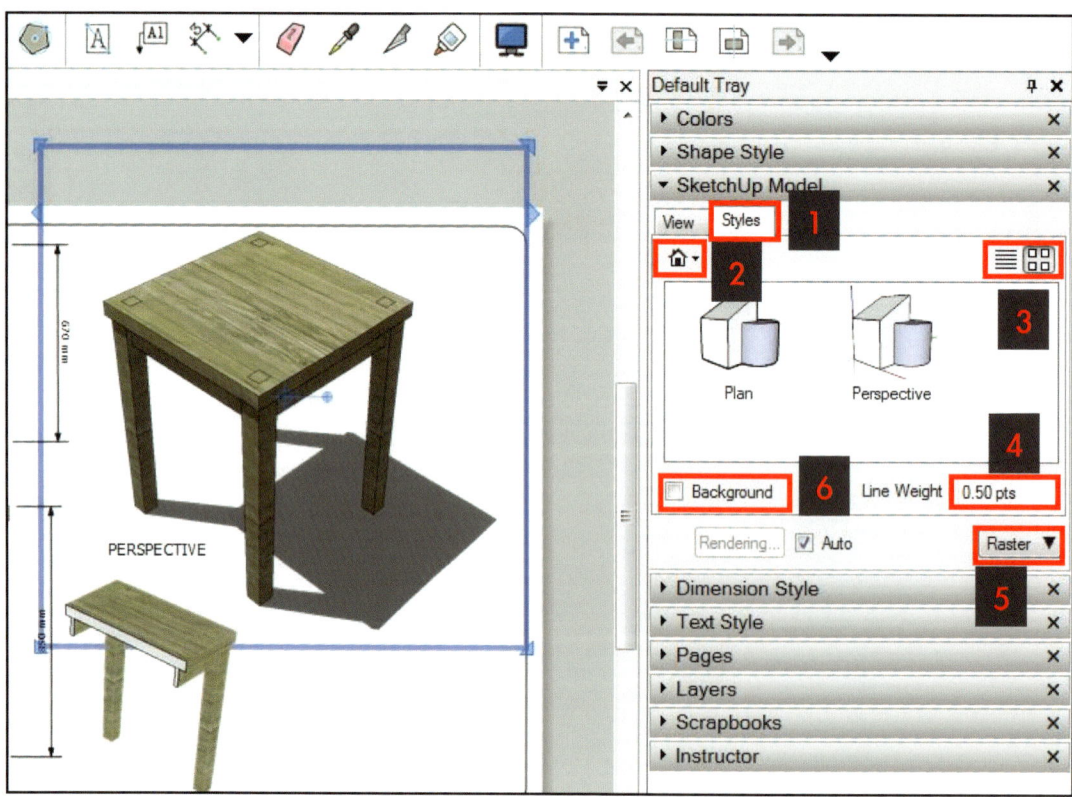

You can export to .dwg/ .dxf (CAD) from either SketchUp or LayOut.

SketchUp2BIM™ Series — Construction Documents in Trimble SketchUp™ Pro by Paul Lee- *A methodology for creating technical 2D drawings*- ©Paul Lee, 2012- www.viewsion.ie

The Result

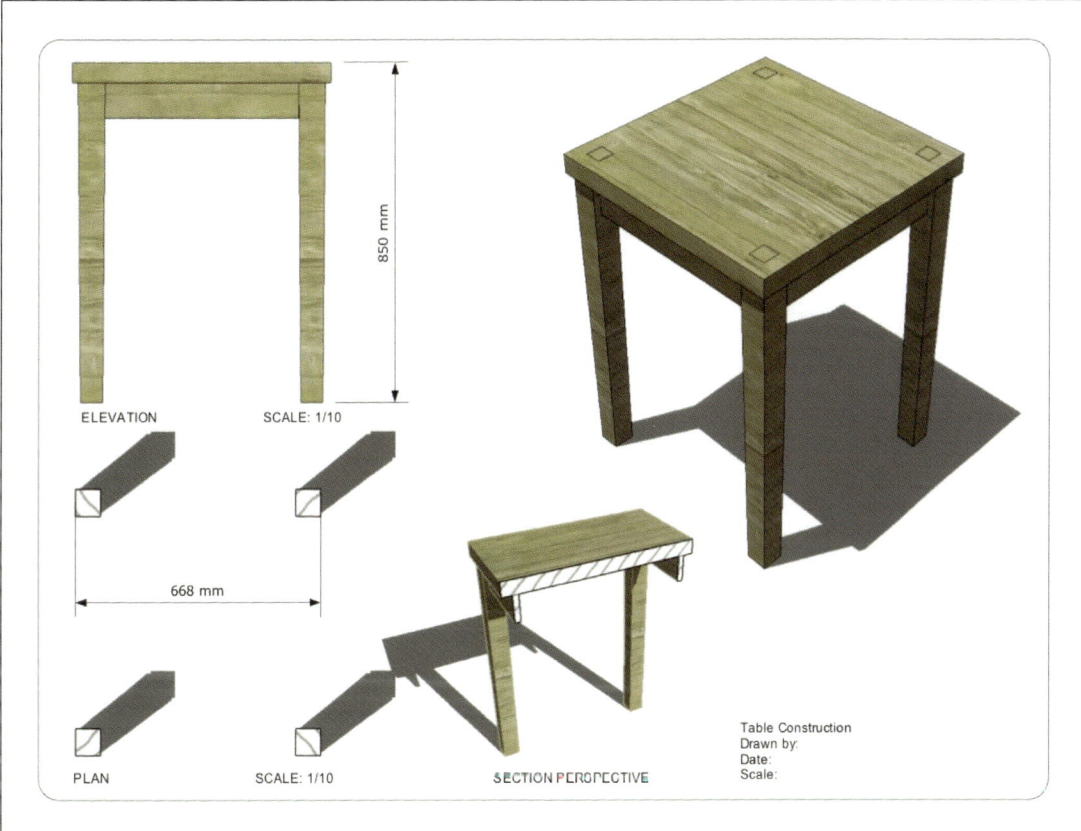

With the skills you have learned here see if you can achieve a drawing similar to the above.

Next: Part 4-
Structuring your model and Recap

Creating Construction Documents in Trimble SketchUp Pro

This section takes about 120 minutes to complete

PART 4 of 4

In this final part of the series, I outline a simple approach that makes it easy to control, edit and display your model. I also provide a link to a structured house file that allows you to examine it's structure and provides a chance to create your own building and drawings.

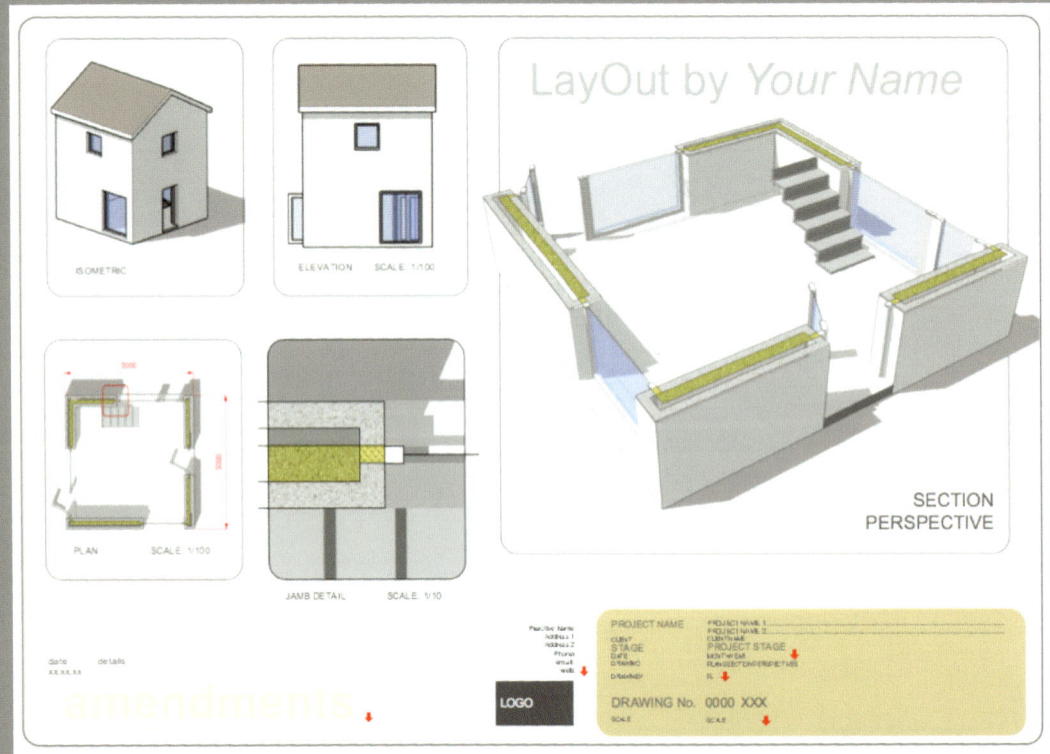

The above drawings represent views of a structured model with 2D construction information.

SketchUp2BIM™ Series — Construction Documents in Trimble SketchUp™ Pro by Paul Lee- *A methodology for creating technical 2D drawings*- ©Paul Lee, 2012- www.viewsion.ie

Basic Model Structure-
Various Groups and their relationships

The best way to group elements so that they can be controlled coherently is to keep every group separate, but for two exceptions: 1. The building model with all it's subcomponents, and 2: External Windows & Doors should be grouped within External Walls. It is really important that groups be formed early on in the modelling process and maintained throughout.

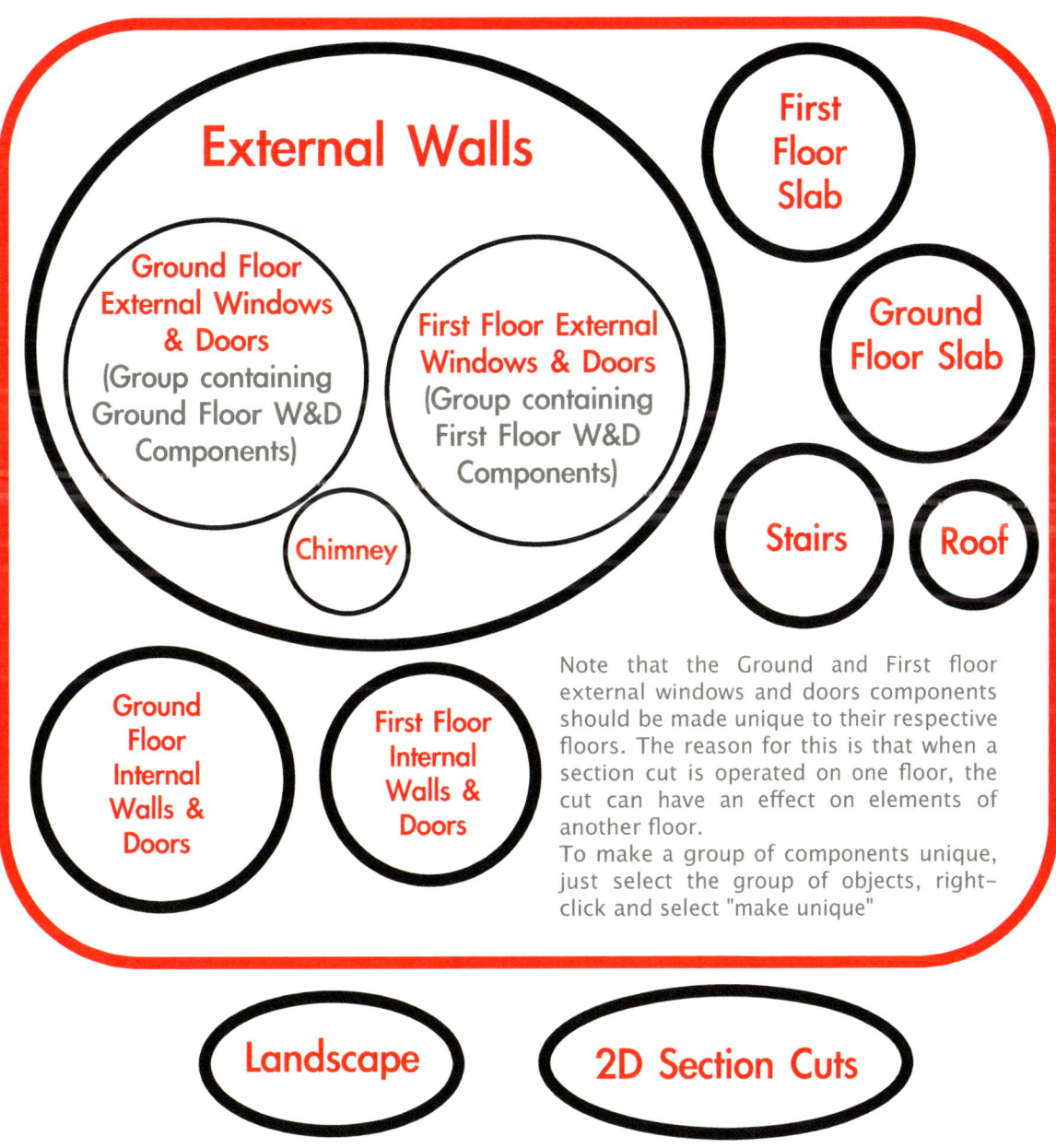

Note that the Ground and First floor external windows and doors components should be made unique to their respective floors. The reason for this is that when a section cut is operated on one floor, the cut can have an effect on elements of another floor.
To make a group of components unique, just select the group of objects, right-click and select "make unique"

Final Steps

1. Download and open the Course Models from:

http://www.viewsion.ie/#!publications

2. Follow the steps in the Scenes embedded in the SketchUp file called: **"Construction Documents in SketchUp Pro by Paul Lee 2012"**

3. Use the file called **"BLANK MODEL"** to follow the steps to achieve the 2D construction information within the model.

4. Next, using the lessons learned from Parts 1 - 3 of this course, see that you are able to complete a titled set of drawings as illustrated in the LayOut File called:

"Construction Drawings in SketchUp Pro by Paul Lee_ Part 4_Target Drawing"

Good Luck!

Thanks for taking this course. Please feel free o email comments, suggestions requests for more information from me: paul@viewsion.ie

Please also visit our website: www.viewsion.ie for more about our training courses or to purchase Pro licenses.

If you would like Viewsion certification for this course you can send me a completed project file of your own using the 2D techniques outlined here. I will review it and see if there's anything you missed. If there is anything you need to do to fix the file, I'll let you know. Then just follow my requirements and I'll certify you. (Note: To request certification costs $99/€99) I will also have a special FAQ on www.sketchucation.com- link through from www.viewsion.ie home page.

Enjoy using SketchUp Pro!

Cheers, Paul Lee

01
02
03
(04)

Thanks to:

Mel, Robert, Anna, Mahon, Teresa for wonderful family support.
Alan, my business partner for his patience.
Shara, Aidan, Chris, John and all the team at SketchUp in Boulder for being cool.
Mike Tadros of Igloo Studios for advice about publishing. Also.
Mike Lucey of SketchUcation for assistance, advice and support.
Bonnie Roskes for inspiration.
Liam Fallon of St. Gerald's College, Castlebar for great advice and support in education.
Camden Palace Hotel, Cork City for providing a workspace and smiling faces.

ISBN-13: 978-1480099012 (CreateSpace-Assigned) ISBN-10: 1480099015

SketchUp2BIM™ Series   Construction Documents in Trimble SketchUp™ Pro by Paul Lee- *A methodology for creating technical 2D drawings*- ©Paul Lee, 2012- www.viewsion.ie

CPSIA information can be obtained
at www.ICGtesting.com
Printed in the USA
LVIC06n1059180214
374179LV00005B/10